Political Ecology of Agriculture

Omar Felipe Giraldo

Political Ecology of Agriculture

Agroecology and Post-Development

Springer

Omar Felipe Giraldo
Conacyt - El Colegio de la Frontera Sur
San Cristóbal de Las Casas, Chiapas, Mexico

Translation from the Spanish language edition: Ecología política de la agricultura: Agroecología y posdesarollo by Omar Felipe Giraldo, © El Colegio de la Frontera Sur, 2018. Original Publication ISBN 978-607-8429-51-6. All rights reserved.

ISBN 978-3-030-11823-5 ISBN 978-3-030-11824-2 (eBook)
https://doi.org/10.1007/978-3-030-11824-2

Library of Congress Control Number: 2019931018

This Springer imprint is published by the registered company Springer Nature Switzerland AG
The registered company address is: Gewerbestrasse 11, 6330 Cham, Switzerland

For my beloved Ingrid

Preface

The subject of agriculture lies within a power-relations struggle that has not been sufficiently addressed by political ecology and environmental thinking. It is true that there are very good case studies from this field of knowledge and excellent papers that touch on the subject of power. Yet, to date, few papers have tried to bring together the theoretical tools that political ecology has borrowed from other disciplines, to unravel and clarify the origins of the discourses, practices, and cultural assumptions, in an effort to explain the conflicts and antagonisms in agriculture. For its part, the field of agroecology has most often addressed the dimension of power in agriculture. However, agroecology is often seen as an alternative route for rural development and sustainable development. We might attribute this to the fact that the conceptual framework of agroecology is not geared to viewing "development" as a cultural project of capitalist modernity. For more than half a century and under the pretext of improving people's lives, proponents of "development" have treated people as part of their political calculations, functionalizing human beings according to the dynamics of accumulation. Hence, there is a need for dialogue between these two fields of study.

In this book, I show that the political discussions of agroecology can find a fruitful space for reflection if such discussions take place in political ecology and post-development theory, and we remove them from the framework of sustainable development or rural development. Although it is true that the problems and approaches of political ecology differ substantially between authors and currents of thought, in this book, I use "criticism of modernity" and "criticism of development" as the foundation from which the relations between capital, culture, and nature are best understood. It is here where the mechanisms for territorial control are set in motion and where corporalities can inhabit different spaces. I believe that the problem of a lack of detailed analysis of the power strategies that are woven into the cultural background of agricultural development and food regimes is that we fail to perceive the subjectivation tactics put in place to subsume the corporalities to the institutional structure that serves the capitalist economic system. Thus, political ecology both becomes the ideal interdisciplinary field for agroecology and considers how the system actively creates the "docile bodies" (Foucault, 2009a) needed for nature to shift from being the living space where we belong as biotic beings to a commodity that is traded in terms of market values (Leff, 2004). Political ecology also analyzes how big capital, in collusion with the state, decides on life in a process that coincides with death (Agamben, 2017).

What agroecology has lacked so far is to lay out the problem of agriculture in terms of discursive procedures, ontological status, and political mechanisms in the historical context that has made it possible for the agroextractive industry to expand globally. This is very different from discerning which public policy is the most favorable or which institutional procedures are the most appropriate to change the agricultural and food regime. Instead, it is a question of defining the agricultural problem in the context of the agricultural strategies that are based on the technical-political and metaphysical rationalities of agricultural geopolitics, as well as the processes that the peasant, indigenous, and other subordinate actors of society carry out in order to re-signify and value their territories in the context of capitalist appropriation.

I argue herein that agroecological social movements should incorporate the criticism of the development project in their critique of agribusiness and the green revolution. Should they choose to do so, they would sharpen their criticism, making it intelligible to themselves and to other social movements that also defend life and territory. Further, the social practices of movements are already providing some very interesting guidelines for transitioning toward post-development and post-extractivism. However, criticism of the development project is much more than semantics. It involves questioning how truth about the agri-food systems is produced while understanding that the political conflict it is waging is, above all, a struggle with the cultural project of modernity and the symbolic order that sustains the metaphysical meanings of extractive agribusiness. I emphasize that the crisis of civilization in which we live is not a problem that derives from having neglected nature in the calculations of development but is rather a symptom of modern dichotomous symbols, which include the separation of subject and object, nature and society, individual and community, mind and body, and reason and emotions. It is from here that individualism, faith in progress, and our anthropocentric self-perception derive, through which we conceive the web of life as vulgar resources available for us to exploit. That is why any alternative to the devastation we are experiencing today cannot start from the same symbols that are taking us to the abyss; on the contrary, it must question its foundations, the structure of meanings on which capitalist modernity is so comfortably based.

Imbibing from political ecology also means accepting that the agroecological struggle cannot be separated from sophisticated regimes of control or the power mechanisms linked to development. Understanding this, on the one hand, will provide greater elements for appreciating the limits of its political action when it is still immersed in the regimes of truth and the hegemonic symbolic structure. It is not easy to escape the power relations that have built the machinery of development for so many years. Thus, the resistance of agroecological social movements and other sectors of society that challenge industrialized agriculture and the globalized food system will have to accept that, if they continue to focus on categorical frameworks of development, they will succumb to the very technologies of power they seek to combat. This will also make it possible for movements to exercise greater caution once they achieve certain conquests in their conflict with the state and multilateral apparatuses, since the danger of ending up granting greater power to the political order from which they want to distance themselves will be more apparent.

The political ecology of agriculture that I present in this book does not endeavor to address all facets of this field of study. It is only an attempt to open an epistemic field for agroecology, which is not sustainable development, nor rural development, but rather political ecology and environmental thinking. I also discuss possible analyses at the intersection of political economy, poststructuralism, phenomenology, complexity, and environmental philosophy. In doing so, I aim, on the one hand, to temper the bias toward productivism and the swarm of concepts derived from the economic rationality inherent in the agricultural sciences. On the other hand, I seek to construct the problem of power of agriculture in the context of antagonisms and conflicts between dissident actors in their struggle to reappropriate nature in a social manner (Leff, 2014). I hope that the analysis herein will also help political ecology and critical thinking in general to visualize the enormous contributions that agroecology's social movements are making to other nonagricultural activities for post-development, post-extractivism, and transition to a world beyond capital.

The book is divided into seven chapters. In the first chapter, I seek to deconstruct some of the meanings of modern thought that give significance to the actions and discourses of agroextractivism in the contemporary world. I start from the principle that we cannot fully understand agribusiness, let alone take away its power, if the roots that sustain productivism are not first comprehended, i.e., the urge to extract the hidden from nature, as if nature were a warehouse of "natural resources" to be extracted, transformed, stored, and distributed as commodities. I try to pose the problem by assuring readers that the political economy of agroextractivism is one more component in the history of metaphysics, that it is consummated in the modern era when human beings conceive of themselves as subject, the rest of the world as object, and pretend to dominate the planet technically. The latter is dramatically expressed today by green revolution technologies, by the landscapes of agribusiness, and by the belief that human history can be explained as a progressive development that moves in the direction of "from less to more."

I begin the second chapter by positing that this modern perspective of conceiving reality generates a type of thought linked to the economy of agroextractivism. This is a way of understanding the world, and ourselves, as if we were governed by the laws of the market, treating everything else as merchandise and conceiving our actions as if they were always motivated by profit. I present a brief environmental history of how this rationality permeates agro-capitalist expansion on a world scale, focusing specifically on the discursive binomial of development and poverty, through which a series of needs formulated in terms of underconsumption were created (Illich, 1996). This discourse "created truth" that created "needs" could be met by bringing more people into the market economy and exposing them to the benefits of technology, which had important implications for the creation of consumers for the agricultural surpluses that plagued the system's postwar operations, for processed food from the food industry, chemical inputs, and agricultural machinery. I conclude the chapter with a data-backed discussion of how ecosystem simplification and pollution created by the model have destroyed the natural sustenance on which capital itself depends in order to continue its tireless expansion. I also show how the system is exploiting this destruction by trying to reconfigure the world agricultural model and open new sources of business by promoting the "green economy" and "sustainable development."

In the third and fourth sections, I turn to the theoretical aspects addressed in the first two chapters to suggest an interpretation of the power strategies of agriculture in the contemporary world. I begin by proposing a dialogue with the most recent contributions of political ecology that employ the notion of "accumulation by dispossession," which David Harvey (2007) and many other authors have constructed by updating the original accumulation described by Marx. The aim is to show that, despite the violent acts of dispossession that have been taking place in recent years, we cannot ignore the fact that territorial control is more powerful the more silent it is, when it works by incorporating, i.e., when done with the consent of the population as a whole. In particular, I focus on the phenomenon of land dispossession in the countries of the Global South, making a phenomenological and constructivist reading of the geographical expansion of agribusiness, based on the principle that to take over land, it is not always necessary to displace its inhabitants physically. Given that it is practically impossible to monopolize all of earth's land and transform it into uniform agro-industrial plantations, capital is making available to plantations the land on which millions of people cultivate and graze indirectly in order to put it at the service of the dynamics of territorial rents. Moreover, it does so when autonomous elements of the technical traditions and peasant economies are intersected by heteronomous elements that change the context in which peasants' learning, fields of enunciation, and daily actions emerge, through the discourses and practices of the new geopolitics of agricultural development.

In the fourth chapter, we continue our dialogue with neo-Marxism to elucidate some strategies of power that are not so evident from the structuralist point of view. The objective is to ask how capital dominates territory, controlling bodies, by redirecting affective relations and sensitivities among peasants and their places of reproduction. In this section, I maintain that territorial control cannot exist unless it is inscribed in the body, in affective feelings, and the sentient horizons of the hegemonized population, creating a frame of reference for what we can really feel. It is a shaping of sensitivities and desires organized by institutions that builds up deterritorialized imaginaries and unravels the social fabric in rural communities. It is not a question of fostering insensitivity but rather of orienting sensitivity by distinguishing what can be felt from what cannot be felt (León, 2011). My hypothesis is that the effectiveness of this conquest of affectivity lies largely in the characteristics of the agrarian aesthetics produced by capital, since it is within such aesthetics that sensitivity arises in one way and not in another. It is in the field of agroextractivism, where daily experience occurs, affections are regulated, and desires and knowledge are administered, that the true regimes of agro-capitalism take on meaning.

The entire first part of the book suggests that the conquest of bodies is the fundamental tool of development, without which the geographical expansion of the agro-industrial locomotive could not be set into motion. Agribusiness actively uses the biopolitics of development to manufacture populations, encouraging people to have a perception of themselves as entities estranged from one another, untied from the land, and self-perceived as merchants who act in accordance to market principles. In other words, as a model of death and desolation, agroextractivism is simultaneously creating forms of human "being" that are congruent with this violent transformation. This is an aspect of the cultural project of capitalist modernity that we cannot ignore in our agroecological struggles, if we intend to dispute, seriously, the hegemony with agroextractivism and the institutional apparatus that supports it and to distance ourselves from the ideological support and the features of power that underlie it.

The fifth chapter focuses on popular organization, which has been growing due to the contradictions of the system itself. Specifically, I discuss the history of the peasant-to-peasant methodology, which I find to be one of the most interesting contributions of agroecology to post-development, given its capacity to recover autonomy, revitalize the network of human relations, and liberate inhibited social powers. Through the exchange of knowledge among peasants, the methodology has demonstrated that it is possible to revive relational wealth, recover the capacity of rural communities to use available resources, and find concrete solutions that are adapted to their cultural experiences and to the ecological particularities of inhabited places while bringing production and consumption systems back under social control. From the perspective of complexity, I discuss the potential of social power when it is self-organized in expansive networks that grow exponentially, allowing knowledge to circulate and new local knowledge to be produced. This is possible because all participants are experimenters and creators of contextually situated knowledge. Peasant-to-peasant teaches post-development that eschews market absolutism or state intervention and widely distributes knowledge through creativity, dialogue, and mutual aid. Beyond discursive matters, I believe that these social agroecological elements are producing some of the most interesting examples for opening cracks in capital. They also point to the fact that we cannot expect change in the capitalist system in the context of state institutions and public policies. It is rather more pragmatic to change capitalism from below by subverting social relations.

The sixth chapter is an epistemological effort aimed at defining some criteria on how technology, economics, and community relations are included in the conditions that make life on the planet possible. Based on the theory of natural drift and autopoiesis by Humberto Maturana and Francisco Varela, as well as on the second law of thermodynamics, I discuss agroecology and agroextractivism from a biophysical perspective and the systemic paradigm of the biological sciences. From Maturana and Varela, we learn that if the basic principle of life is that ecosystems are organized autonomously in a nonlinear process, then any intervention therein cannot be linear. This means that everything is permitted in human creativity, except for nature's only restriction that must be respected: no action should impede the integrity of the substrate that the agroecosystem needs to survive. Human capacity for intervention and technical innovation is limitless, as long as we follow the rule of not impeding the reproduction of the cycles that make life possible. This also means adapting to the laws of thermodynamics. Based on an understanding of energy flows, we discuss how different people managed over the past 10,000 years to live and survive on agriculture in their territories without depredating the environment. We might explain this by noting that their ways of transforming the ecosystem were adapted to the negentropic organization of living matter, which included living in coexistence with biodiversity and ecosystem loops. These biophysical principles imply that there is no room for the monotonous repetition of universal techniques; rather, there is infinite room for cultural creativity so that social organization can be integrated into the negentropic and autopoietic process that constitutes the natural order.

The final chapter examines the future of industrial agriculture given the decline of the fossil fuel era and the likely collapse of industrial-based civilization, as the material base on which the entire system depends disappears. This section discusses that a perceived and increasingly artificial, hyper-technologized, gray, and disarticulated future—like the images of Hollywood cinema—responds to linear scenarios, based on the idea of scientific and technical progress and the preconception of the city as the ultimate goal. However, given the impossibility of continuing to feed energy into the dynamics of accumulation, we face a critical disruptive scenario, in which the system must self-organize in another way. If we take seriously the inevitable depletion of the energy and material sources that underpinned the growth of industrial civilization and capitalism, then we can imagine many other possibilities for the future: deindustrialized, de-urbanized, and greener, smaller-scale societies, with simpler technologies, and a massive return to rural settlements. Without trying to prophesize, I argue that in any potential scenarios, agroecology will accompany our transition to other forms of civilization. In fact, I think this is an opportunity to dream of other landscapes with forests integrated into an agroecological rhizome and with human populations living in their interior. In any case, other future possibilities different from the imaginaries of progressive artificialness would require not only a change in the technical and political-economic platform but also a profound ontological and spiritual change.

Finally, I believe that the ecological agriculture and its landscape transformations have profound ontological implications. The landscapes of diversity, like those that the agroecological project aspires to achieve, could be part of the backdrop where the spiritual transformations we need to return to an understanding of ourselves as interdependent beings, hyper-related, and belonging to vital networks take place. Ontological changes—from a split "being" to an interconnected one—will not occur away from where we establish residence, since they happen within the habitats that we forge with our actions. Far from crying, "Do not touch me," nature is shouting for the transformation of the ecosystem without breaking the biotic equilibrium or transgressing the natural niches of plants and animals and their symbiotic relationships. It is also a cry for the compatibility of our symbolic order, to understand ourselves as a product of the interrelations, interdependencies, and complementarities that we inhabit and that inhabit us.

I would like to end this presentation by briefly explaining the context in which this book emerged, which will also give me the opportunity to extend my gratitude to a number of people. *Political Ecology of Agriculture* is the result of 5 years of research on post-development agriculture. Although some of the ideas discussed here were previously published elsewhere, this book endeavors to present a panoramic view of these analyses and other unpublished ideas that close a productive period of reflection. The opinions presented here were discussed in various fora, and many people contributed in various ways to their preparation.

The book began taking shape during a research internship I did at the Academic Working Group on Environmental Thinking at the National University of Colombia, during the second semester of 2012. There, I had the immense fortune to share ideas with the gifted philosopher Ana Patricia Noguera and the team of dreamers who accompany her. To her I owe immense thanks for her invaluable friendship and for welcoming me to this academic environment so that I could develop some of the concerns I had then about agroecology in an aesthetic and phenomenological perspective. That academic experience left an indelible mark on me that is reflected herein.

This book began its journey in August 2013, months after obtaining my doctoral degree, when I enrolled in a postdoctoral course at the Institute of Social Research of the National Autonomous University of Mexico (UNAM). There, I was advised by the great Mexican epistemologist and environmentalist Enrique Leff. Thanks to his guidance and a grant from the UNAM, I was able to carry out research on territorial control mechanisms within the framework of land dispossession in Latin America. To Enrique Leff I extend all my gratitude and admiration. Connoisseurs of his monumental work will detect a small part of his thinking reflected in this book.

In September 2014, I was able to continue my research when I obtained an academic position as professor, with support from the National Council of Science and Technology of Mexico, at the El Colegio de la Frontera Sur—ECOSUR—in San Cristóbal de las Casas, Chiapas. Thanks to Helda Morales, who invited me to participate as a researcher in a project on "agroecology massification," I joined a gifted group of researchers that includes Bruce Ferguson, Peter Rosset, Mateo Mier y Terán, and Miriam Aldasoro and a dynamic group of our master's and doctoral students. By participating in this collective, I developed the ideas related to the agroecological social movements mentioned in the second part of the book. I want to express my infinite gratitude to Helda Morales and the entire group that has become part of my family.

Peter Rosset merits a special mention. His long experience as a technician at La Vía Campesina and as an eminent theoretician of political agroecology was fundamental to providing detailed information on the experiences of many organizations discussed in this book. Together with my esteemed colleague and friend Mateo Mier y Terán, and the rest of the academic group, we have been able to discuss, expand, and reflect on numerous topics that appear in the chapters herein. Without Peter, I could hardly have risked writing details regarding the peasant movements, the peasant-to-peasant methodology, and, in general, the social dynamics of agroecology on a global scale.

I am also indebted to my dear postgraduate students who participated in the seminars in Agroecology and Society, Environmental Thought and Political Ecology, and Political Agroecology at ECOSUR, as well as in the Theories of Rural Development seminar at the National University of Costa Rica. Many of the arguments were extensively enriched during the discussions we held during these courses. The doctoral seminar in Agroecology and Society of which I am co-coordinator, together with Fabien Charbonnier and Mateo Mier y Terán, has been an especially important space to learn about approaches and discussions about which I was unaware and are now included in this book.

A very special moment that would mark the preparation of the book was a seminar on Ivan Illich's thought organized by Susan Street, who is the Jorge Alonso Chair of the Center for Research and Advanced Studies in Social Anthropology (CIESAS) in the city of Guadalajara. It was an unsurpassed opportunity to explore Illich's work, discuss my ideas with the lucid Gustavo Esteva, and begin a beautiful friendship with Astrid Pinto Durán, an esteemed colleague within the Seminar on Culture and Climate Change. I had an important opportunity to learn firsthand about the Cuban experience of the National Association of Small Farmers (ANAP) in 2015 and 2017 and to initiate action-research work with peasant organizations and other groups that promote agroecology in Chiapas. Furthermore, my life in San Cristóbal de las Casas, in the context of the Zapatista movement and current discussions of the Indigenous Government Council of the CNI, has left its mark on my political and epistemic orientation.

I acknowledge my beloved wife Ingrid Toro for bringing to my attention the importance of affectivity and empathy within the power exercises of development. The chapter on this subject is the result of many dialogues that we had, and it is here that her own words have been inscribed. I am also grateful for the work of my beloved father, teacher, and writer José Omar Giraldo, for helping to style edit this book and for pointing out aspects I had failed to take into account in the original draft. I am also indebted to my friend and writer Andrés Felipe Escovar and Pierre Madelin, Valentín Val, and Fabien Charbonnier, who made pertinent observations that enriched and improved the manuscript.

Finally, my gratitude goes out to my friends Ricardo Andrés Lozada, Sergio Amorocho, Jairo Andrés Beltrán, Julián Toro, Andrea Moreno, Carla Zamora, Renzo D´Alessandro, Julián Pérez, Vera Camacho, and Alberto Vallejo, my beloved mother Rosita Palacio, my sister María Elena Giraldo, my niece Julieta, my wonderful in-laws Judith Velosa and Erdulfo Toro, and all those people present in my daily life who I cannot name here. We too are inter-beings formed by our friends, our family, and the beings we love: they are part of us in such an inseparable way that we find it difficult to determine where their influence begins and ends.

San Cristóbal de Las Casas, Mexico Omar Felipe Giraldo

References

Agamben, G. (2017). *Homo sacer. El poder soberano y la nuda vida I*. San Cristóbal de las Casas, Mexico: CIDECI-UNITIERRA.

Foucault, M. (2009a). *El orden del discurso*. Mexico: Tusquets Editores.

Harvey, D. (2007). *El nuevo imperialismo*. Madrid, Spain: Ediciones Akal.

Illich, I. (1996). Necesidades. In W. Sach (Ed.), *Diccionario del desarrollo. Una guía del conocimiento como poder*. Lima, Peru: PRATEC.

Leff, E. (2004). *Racionalidad ambiental. La reapropiación social de la naturaleza*. Mexico: Siglo XXI Editores.

Leff, E. (2014). *La apuesta por la vida. Imaginación sociológica e imaginarios sociales en los territorios ambientales del sur*. Mexico: Siglo XXI Editores.

León, E. (2011). *El monstro en el otro. Sensibilidad y coexistencia humana*. Madrid, Spain: Sequitur.

Contents

1 Agroextractivism: The Desert Is Growing! 1
Extractivism: Bringing Out the Unseen 2
Metaphysical Thinking and the Green Revolution 5
Progress and Development: Temporary Certainties of Agro-Capitalism 11

2 The Economic Rationality of Agroextractivism 17
Economic Rationality and Agriculture 18
Agricultural Overaccumulation and the Development Project 23
The Environmental Consequences of the Commodification of Agriculture and the Restructuring of Agro-capitalism 28

3 Territorial Control and Geographical Expansion of Agribusiness 35
Crisis of Capitalism and Violent Dispossession of the Land 36
Farmers in the Global South: Ontological and Epistemic Colonization 42
In Situ Deterritorialization of Peoples 46
Inclusion and Removal of Autonomy 47
Control of Enunciation Fields 53
Agribusiness and Territorial Rent 56

4 The Government of Affections 61
Creation of Linkages with the Place and the Political Administration of Affections 62
Constructing a Shortage and Producing Desire 67
Esthetics of Agro-Industrial Progress 70

5 Agroecology in Post-development 75
Social Processes of Agroecology 78
Regeneration of Community Spaces and Socialization of Vernacular Wisdom 85
Agroecology and Cracking Away at Capital 90

6 Reinhabiting the Earth's Crust Through Agroecology 97
Agriculture, Coevolution, and Natural Drift 98
What Is Life? Autopoiesis and Agroecology 103
The Second Law of Thermodynamics in Agroecology 106
Creativity: Technique and Technology 112

7 The Future, Behind Us 117
Possibilities of Worlds Turning Green Again 121
Ontological Transformations, Spiritual Transformations 127

References 135

Index 145

About the Author

Omar Felipe Giraldo holds a PhD in Agrarian Sciences from the Autonomous University of Chapingo. His thesis was awarded the Arturo Fregoso Urbina Prize for the best doctoral thesis of 2013 and an honorable mention for the Jorge Alonso Prize for the best doctoral thesis in social sciences in Mexico, 2013, awarded by the Center for Research and Advanced Studies in Social Anthropology (CIESAS) and the University of Guadalajara. He served as advisor-consultant to the United Nations in Colombia for several years. He has worked as a postdoctoral researcher and professor of sociology at the National Autonomous University of Mexico, postgraduate professor at the National University of Costa Rica, and guest professor at the National University of Colombia. For several years, he has worked in the field of research in environmental thought, political ecology, post-development, and agroecology. He is currently a Conacyt researcher affiliated to *El Colegio de la Frontera Sur* (ECOSUR) and is part of the National System of Researchers of Mexico. He is the author of numerous articles and the book *Utopias in the Age of Survival: An Interpretation of Good Living* (2014) [*Utopías en la era de la supervivencia. Una interpretación del Buen Vivir*].

Chapter 1
Agroextractivism: The Desert Is Growing!

> This civilization advances in love with war; in love with the desert that it creates in its wake.
> Jaime Pineda

Understanding the foundation on which the practices and beliefs of extractive agribusiness are based is a task that involves deconstructing the culture that supports it. This essential task is not usually considered of interest by those who work in agroecological studies, since there is a deep-rooted certainty that the problems of the model can be solved by making technical and sociopolitical changes, without taking into account the structure of meanings and the meanings of the inherited culture. As we will see, this idea is at the very heart of the problem. In a very different way, I will argue that the agribusiness that today seeks to establish itself in the croplands of the entire world is imprinted with the mark of inherited Western culture. For this reason, the critique of agroextractivism undertaken throughout the book begins with the history of thought that grounds the logic and certainties of a destructive activity that is leading us to the abyss. Undoubtedly, in our times of civilizational crisis, changes of all kinds are unavoidable, but we should emphasize that this crisis—of which the agri-food system is an integral part—will not be overcome if we do not first reflect deeply on the roots of our civilization.

We should clarify that the critique herein is not intended to cover all cultures on this Earth, only a very specific culture whose origin has a very precise historical and geographical location. For pragmatic reasons we will call this culture "Western," although the truth is that today we cannot limit it to a specific territorial space that can be called "Western." African, Asian, and Latin American countries, abetted by their governments, can often self-inflict the negative consequences of Western culture more dramatically and with greater harm than the European nations where the birth and consummation of that culture took place. Extractivist policies are a good example of this and are part of what Latin American analysts have called "the coloniality of power" (Quijano, 2000). Beyond the historical questioning of how

O. F. Giraldo, *Political Ecology of Agriculture*,
https://doi.org/10.1007/978-3-030-11824-2_1

this culture was inherited in so many spaces around the world, in this chapter we seek to understand its meanings and the way in which these meanings guide and orient the actions and discourses of agroextractivism.

In this chapter, we will concentrate on an aspect that philosophy calls "metaphysics," which, as we shall see, determines the foundations of modern thought to this day. We will deconstruct some assumptions of Western philosophy that arise in Greek civilization and modernity, which are fundamental to understanding the representations that give firm ground to extractivism in the contemporary world. We will begin with the notion that constitutes the basis of productivism and on which the narratives and actions of agribusiness are based, which we trace from the beginning of today's so-called "Western" thought.

Extractivism: Bringing Out the Unseen

> I used to be able to rejoice over a new truth, a better view of what is above us and around us, but now I fear that I shall end like old Tantalus, who received more from the gods than he could digest.
>
> Friedrich Hölderlin

Heraclitus is known as "the dark one," more because of the difficulty of understanding his approach than because of the meaning of the term itself; that is, he conceived of "darkness"as essential to understanding the totality of what exists. This dimension of darkness, so important in Heraclitus's proposal, was abandoned by Western thought following the route charted by a contemporary of his, the philosopher Parmenides.

Heraclitus started from the idea that nature is always in constant movement; it flows incessantly, alternating between day and night, heat and cold, light and darkness, and summer and winter. But not as mutually exclusive opposites. Rather, he understood them as complementary elements (Schüssler, 1998). Interestingly, the notion of complementarity, so ephemeral in Western thought, played a fundamental role in Chinese civilization 2500 years ago. This way of understanding the world, in Taoist wisdom, is represented by the well-known figure of *yin* and *yang*, which symbolizes how opposites maintain a complementary relationship, where one is always in the other. Even today, many of the original peoples of the American continent think of day and night, sky and earth, sun and moon, light and dark, truth and falsehood, and masculine and feminine as necessary complements, since they believe that opposites have always been bound together in a relationship of unbreakable reciprocity (Estermann, 1998).

Similarly, Heraclitus sees a relationship of unity in opposites, i.e., when one appears the other disappears. Day gives way to night to the same extent as night retreats from the advancing day. Thus, opposites have always been linked in this relationship of indissoluble unity: "Day and night are one," Heraclitus said. The process can be described as an associating movement, i.e., the rise of one implies the decline of the other. In itself, the day cannot be what it "is" if there is not, within it, darkness in search of light. The lighter the day, the deeper the darkness of the night. The more opposites are contrary to each other, the more they reinforce each other (Schüssler, 1998).

According to Heraclitus, there is always something hidden in the unity of opposites, because the hidden is a constituent part of that unity: "all is one," he maintained. In the presence of day, night is hidden, just as in the presence of night, day is hidden. The key issue is to understand that nature moves in a relationship of mutual opposition. Yet, this is not a characteristic that can be considered negative in any way. On the contrary, the dimension of darkness watches and protects while offering recollection and rest so that the day may dawn each morning. To the question of how the magic spark of life is possible, Heraclitus's answer could be summed up as "in the incessant disappearance." Only by hiding can one guarantee the seclusion and rest that life requires to continue its perpetual movement (Schüssler, 1998).

That is why Heraclitus asserts, "nature likes to hide." However, Heraclitus is not considered the founder of Western philosophy, an accolade garnered by his contemporary Parmenides, who traced out a very different path, which in the end would chart the course of Western thought. For Parmenides, movement and fluidity—as Heraclitus thought—is an illusory appreciation. The existing is immovable, constant, and immutable, he believed. Although for Parmenides it would seem that the day is present and to that extent "is" and that it seems to disappear when night falls—creating the idea that it no longer "is"—in reality the absent has not ceased to "be." Certainly, the present day passes into absence, but absence cannot be considered as a "nothing," as a "not being"; on the contrary, it is something that "is." Parmenides assures us that the absent as such has a presence. Even the absent is always present. That is why "being" itself never ceases to "be" (Schüssler, 1998).[1]

Yet what do these entangled philosophical ruminations have to do with the practices and discourses of agroextractivism? The answer is that Parmenides opened a path in Western culture on which we continue to tread. I refer to the idea of the predominance of the category of presence, of the diurnal dimension that corresponds to excessive productivity, and of the forgetfulness of the nocturnal dimension proposed by Heraclitus, i.e., that which offers rest, recollection, and inaction, so that day may dawn again. Agroextractivism's interventions have their origins in this insatiable search to uncover all of Earth's elements, extract, and make them available, converting them into "useful" resources for economic accumulation and capital valorization.

There are many examples of agroextractivism, but examining the case of soy is sufficient. To produce 1 ton of soy in the "green deserts" that have rapidly territorialized the fields of the Global South since the dawn of the millennium, "16 kg of calcium, 9 kg of magnesium, 7 kg of sulphur, 8 kg of phosphorus, 33 kg of potassium, and 80 kg of nitrogen" must be extracted (Anino and Mercante, 2009, p. 82). These chemical elements are not replenished in the soil, and thus, it degrades, rapidly undermining the reproduction of life. Here, an arrogant civilization eagerly strips nature until it extracts and exploits the last bastion of minerals, in order for this brand of agriculture to continue having a permanent, growing presence while

[1] The reflection is based on the poem *On Nature*.

denying the seclusion, the rest, and the non-doing of the night as a counterpart to the daytime dimension. "The desert grows," Nietszche (2000, p. 731) wrote in a prescient statement regarding planetary extinction, because productivity without replenishment is impossible without devastating the earth, without drying up the nourishing soil, and without destroying the fabric of life.

Heraclitus said, "Nature loves to hide." Only when it rests is it possible for nature to regenerate itself constantly. Of course, proponents of extractivist technology fail to understand this, and in their eagerness to undermine the soil and subsoil in a bid to uncover what nature has hidden in the depths of the geological strata in the form of oil, gas, coal, and minerals, they inhibit vital forces by suppressing them. The output of extractivism "brings out the unseen," as Heidegger (1994a) put it. This activity seeks to bring to light; it endeavors to make hidden nature, protected in darkness and silence, appear as a commodity, as resources made available and traded on world markets. This is biotechnology at the service of corporate capital, "bringing forth" nucleic acids that are hidden deep in the cells to manipulate them at will, so that they emerge resistant to the chemical poisons that capital itself produces.

The biotechnology of agroextractivism unlocks the most intimate secrets of life and makes them manifest as a presence; that is, it makes them appear, leaving behind the state of concealment in which they were kept (Heidegger, 1994b). As Heidegger (1994a) asserts, modern technology is nothing more than a mode of "revealing." It is a type of challenge that requires nature to supply "natural resources" in order to be extracted, transformed, stored, and distributed as stocks. For thousands of years, plants have hidden carbon dioxide gas underground in the form of oil and coal, elements that are rapidly emerging from their hiding places as mineral deposits for consumption by an energy-hungry industrial civilization. For eons, the soil stored minerals that are now plundered through multifunctional monocultures intended for fattening stabled animals, malnourishing human beings, or feeding cars.

Extractivism is a way of making present all that was hidden in the calm of the hideaway, in the nocturnal dimension of waiting, abandonment, and renunciation. It brings out what nature had hidden, as if it were a storehouse of stocks and a resource warehouse at the service of economic accumulation. Extractivism pursues nature as a set of available reserves where the hidden is forced to appear. In this realm of abundance and excess, the key word is productivity. The more efficient and effective exploitation is, the more life is extracted and transformed into energy, and consequently, capital accumulation continues. Productivity is the touchstone to which all political alternatives that seek to challenge capitalism's seemingly unshakeable truths must refer. Any ensuing ideological dispute triggers a struggle over the means, but not a conflict over meaning. On both the left and right of the political spectrum, the question is how to make productive forces grow and thus generate great wealth, without questioning how wealth grows.

Productivity is an irrefutable answer, since Western culture, from its origins in Parmenides's thought, ignored the dimension of frugality, serenity, and nocturnal calmness. All scientific-technical knowledge strives to "present" what is extracted from hiding places, in order to produce more with less capital investment.

Agribusiness is nourished by these productivist meanings, which in the short term help it to increase its yields, but at the cost of breaking the relationship between crops and ecological systems, and genetically eroding life that has been naturally adapted for centuries in localized niches. With its unnatural intervention, the nutrient cycles and energy recirculation of food chains are abruptly replaced by linear flows based on urban industrial production. This process impoverishes soil fertility, pollutes water, saturates the atmosphere with polluting gas emissions, speeds the clear cutting of forests, and devastates biodiversity, leading to permanent destruction and the ongoing expansion of the desert (Nietszche, 2000).

This route of planetary extinction caused by impoverishment and loss of land can best be understood by examining the foundations of modern Western culture. Thus, we continue our analysis of Western thought to comprehend the environmental degradation of the contemporary world.

Metaphysical Thinking and the Green Revolution

The word "metaphysics" literally means "beyond physics" or "beyond nature." It denotes a way of understanding the world whereby nature is at our service, always subordinated to our wants. It corresponds to a thought that moves away from the roots of the earth. It cuts off the ties of the substratum to which we belong as biological beings and we embark on a suicidal adventure premised on the belief that we no longer belong to the earth (Nietszche, 1999). Metaphysics lends backing to the certainties behind the modern way we relate to nature and characterizes the economic rationality found in discourses and practices of extractive agribusiness. Metaphysics does not begin with modernity, but we would do well to comprehend it because modern thought is based on its structure of meanings. To understand it, we turn to the philosophy of Plato, who outlined in *The Republic* "the metaphor of the cave," an excellent example of the foundations on which the vast array of symbols of Western culture was built.

The metaphor described by Plato (Platón, 1958) is as follows. Underneath the earth is a cavern that has an exit onto the light of day. Inside live men whose feet and necks have been bound with chains from birth, such that they can only direct their gaze in one direction. In their lifetime, they have known nothing but shadows reflected by a bonfire behind them. One day, one of the men escapes his chains, and as he turns his gaze to the blaze, he realizes that it is none other than the light of the fire that glows behind him that causes the shadows he has seen since childhood. The prisoner, now liberated, makes his way out of the cave, where little by little, he begins to comprehend the existence of the water, the trees, and all the beings of a new world that had been hidden from him. After this amazing discovery, the man raises his head toward the sky and observes for the first time the radiant sun that dazzles his eyes. His new insights lead him to conclude that the light that shines in the immensity of the sky is the source of illumination around him, even the shadows he saw inside the cavern. He thus concludes that the sun is the supreme source of everything, penetrating and governing all things. The man, once provided with the truth, far from wanting to return to the cave, will pity his former fellow prisoners, who continue living in the shadows.

Plato explains the metaphor by stating that the cave is the image of a sensitive world, that is, the world we perceive with our senses, while the outside world illuminated by the sun is the suprasensitive or intelligible world, that is, the things that are only accessible through the intellect, "the world of ideas" he called it. That world, in turn, is illuminated by the supreme idea, the idea of ideas, which is the source of all light and is metaphorically represented by the shining sun that blinds the eyes of the liberated prisoner. The ascent of man from the cavern to the light of day is the image of the soul, which ascends from the sensitive world to the suprasensitive, even to the supreme idea, source of all light and which illuminates the whole world. However, the explanation for understanding the meaning of the metaphysics that interests us here is that for Plato the cave is man's sojourn on earth. The underground room, in which these people live, anchored to the ground, must be abandoned in order to rise to the light of ideas where the real world is, and not that apparent and deceptive environment of cave shadows (Schüssler, 1998).

Metaphysics, understood as the refusal of the earth, the attachment to the light of reason, and the belief in a world of the "beyond" as the only true one, begins properly with Plato. Later, it will penetrate deeply into the Western thought with the Neoplatonists, Christianity, and modernity until it completely invades contemporaneity, currently given dimension as the environmental crisis. The metaphor of the cave endeavors to show us how the sensitive world, the world of the "here," is imperfect; it is only a meager copy of a far superior world: the realm of ideas, which with the advent of modernity becomes the empire of reason. True knowledge corresponds to the sphere of the "beyond," and things perceptible by the senses are only of value insofar as they are influenced by the idea of ideas, the supreme good symbolized by the sun, the source of all light. Hence, human life only acquires meaning when the spirit leaves the realm of the senses and communicates with the supersensitive world of thought, as Descartes would demonstrate 20 centuries later through his systematization of the scientific method. Neoplatonism—whose main representative was Plotinus—placed the supreme idea in the absolute, all-embracing spirit of God, from whose perfect fullness one descended "gradually to the minuscule world of matter" (Randall, 1952, p. 51). Subsequently, for Christianity—which has been fundamentally Platonic in its philosophy and values—the meaning of life would be to direct the vision upward to seek eternity in the divinity far removed from the materiality of this shadow-ruled world. The earthly soil where human beings live must be rejected as evil and degraded—a wicked valley of tears—and a higher world, that of God, must be pursued instead, for human destiny is not found on earth, but belongs to another kingdom (Randall, 1952).

Yet, it is not in Christianity proper, but in modernity, that the metaphysics initiated by Plato is consolidated. During the sixteenth century, a new era began in Europe, which would eventually envelope the world. The certainty that during the Middle Ages was based on faith would henceforth be based on reason. In modernity, the support of certainties would rest on the truths provided by science and technology, and the sensitive nature described by Plato would submit to the laws of reason—the world of ideas—to later become resignified in order to transform it into merchandise, into resources available for the accumulation of capital. With modernity, the world begins to be perceived as an object in unison with a process that leads human beings to become subjects. This identification initiates humans' manner of being, by which they represent themselves as the center of the totality of the existing thing, and thus perceives everything nonhuman as a function of its utility, even as they value it according to its possibilities of being exploited (del Moral, 2004; Heidegger, 1996). Metaphysics as conceived in the modern era, which finally becomes the foundation underlying Western culture, provides an ordering that no longer requires philosophy or Plato. The meanings with which modern humans apprehend things are so firmly rooted in the metaphysical understanding of the world that we no longer need to refer to philosophy to understand a civilization that today has become catastrophic.

The current environmental crisis is a consequence of our pretensions of technically dominating the living planet we inhabit, going back to the fundamental rupture by which Western thought was separated from the immanence of earthly space. In the modern era, this rupture has brought the unchecked power of calculation, planning, control, manipulation, and domination of nature. The Green Revolution, which began in the 1960s, is an excellent example of this particular way in which hegemonic culture relates to the ecosystem order. This revolution refers to a package of technologies that included the introduction of high-yield plant varieties, irrigation or controlled water supply, the use of chemical fertilizers and pesticides, the massive use of agricultural machinery, as well as the "improvement" of genetics, nutrition, and animal health. Its objective is the modification of the environment "in such a way as to create conditions for agriculture and livestock farming that are more suitable than those offered by nature itself," says a document of the Food and Agriculture Organization of the United Nations (FAO) (1996). In its Neoplatonic and anthropocentric way, the document exemplifies how nature is apprehended, as if the world were imperfect and needed to be improved. So if the climate is dry—the report continues—"irrigation is used; if soil fertility is low, fertilizers are applied; if pests and weeds invade crops, they are sprayed; if diseases threaten livestock, vaccines and medicines are administered; or, if more energy is needed to plow the land, mechanization and the use of fossil fuels are used."

The zeal of agribusiness to subdue nature and modify it in accordance with the requirements of capital accumulation to maximize returns has led the apparent effectiveness and utility of Green Revolution technology to its antithesis: the devastation of the Earth, the desertification of life forces, and the progressive collapse of the ecological conditions needed for life to survive. The metaphysical dream of rejecting the relationship with the earth and ascending in the light of ideas has led to the tragedy of Western civilization (Ángel, 2002), i.e., our modern aspiration to subdue nature, which has done nothing but produce desiccation and debilitate life's interconnections. The romantic painter Francisco de Goya foresaw the collapse of the world caused by metaphysics when he titled a premonitory engraving "the dream of reason produces monsters" (Pineda, 2016). Today this omen seems less about reason focused on meditative and serene thoughts about purpose, and more about the dream of instrumental reason focused on finding the most suitable means of looting, extracting, and dominating without any concerns about goals (Horkheimmer, 2002), which, as in Mary Shelley's *Frankenstein*, have created lingering monstrosities (Noguera, 2004).

Agro-capitalism, addicted to the excessive productivity of commodities, has depoetized the world, and has turned nature into a vulgar reserve of stocks that must first be scientifically explained and understood (Noguera, 2004), and then intervened, manipulated, and directed by agro-technology, making goods appear on a productive assembly line pushed to its maximum use and exploitive capacity. When the metaphysical is expressed in agro-capitalism, land becomes a thing, an object at the service of corporate ambitions. We have abandoned the land we belong to and have climbed to the limits of a calculating reason to consume nature avidly and without moderation. Scientific research subordinated to big capital finds "more violent methods to transform the celestial forces into energy" (Janke, 1988, p. 49), because the earth, for business capital, is not a sacred dwelling, but a factory that can be calculated, planned, and directed, thanks to the manipulations of biotechnological engineering and strategic management. For agribusiness, in a way similar to what Galileo believed, there is no doubt that nature is an open book that can be read in mathematical language (Pardo, 1991), so that it can be understood and manipulated in terms of efficient management and administration to reap the most gains with the least economic investment. Metaphysical Western culture was completely oriented toward the light of ideas; agroextractive reasoning is focused on calculating reason.

At the beginning of the modern era, scientific-technical domination of all sectors of the world was given free rein, and this ascendancy is now consummated in an era characterized by economic globalization and the planetary hegemony of the metaphysical rationality initiated by Plato. Human beings conceive of themselves as those who "give the measure to everything and set all standards" (Heidegger, 1996, p. 77). They express themselves faithfully in the transformation of motley natural landscapes into uniformed sowings of genetically modified plant varieties; the conversion of tangled tropical forests into homogeneous and extensive pastures; and the confinement of enormous animal populations into stables, sheds, and pigsties. Nature as object has caused the human animal, and our biological counterparts, to wander now through a geographically expanding desert (Heidegger, 1994b).

Agricultural technopower mistreats the land, changing it artificially and abusively. Instead of being inserted in ecosystem loops, it forces the land beyond what is naturally possible (Heidegger, 1994b), since the aim of agro-industrial business is not to make a home on the land, but to discipline biodiversity, selecting what is useful for its exchange value and eliminating what is useless for capital accumulation. Exuberant nature is transformed according to the industrial image of industry (Shiva, 2007), so that agrarian landscapes become homogeneous spaces with a marked predominance of the straight line and quadrangular figures typical of Euclidean geometry. Industrialized rurality becomes the aesthetic image of urban architecture (Noguera, 2004), so that chaotic ecosystems are mutilated and geometrically ordered by mechanical devices that flatten and delimit them and function in clear arable portions for the control and securing of hard currency in keeping with the designs of economic valorization and greed of capitalist civilization (Giraldo, 2013).

Agro-biotechnological activity forces plants, animals, and people to establish a regulated and disciplined group of living beings in order to impose patterns on nature that originate in the factory (Shiva, 2007). Now it exercises power over life, not only by separating and displacing species adapted for centuries in ecological niches but also by manipulating and modifying the genetic code. Biotechnological science at the service of corporate capital intervenes and transforms genes so that a few transnationals can exercise power over the essence of life, appropriating it as if the nucleotide chains were now of their making. A good example of this are the genes used to sterilize the second generation of cotton, rice, wheat, and soy seeds, thus eliminating their reproductive potential and forcing farmers to buy seeds from these same agro-biotech companies during the sowing season (Díaz, 2011). Technological biological power, capable of manipulating nucleic acids, penetrates into the materiality of life and its most intimate secrets, demonstrating its capacity to dominate nature. It undertakes, in the most sophisticated way, the metaphysical dreams of putting reason at the service of the scientific-technical exploitation of the world, at last becoming "masters and possessors of nature" as Descartes fantasized (2008, p. 38) in his *Discourse on the Method*.

As Enrique Leff (2004) asserts, knowledge not only names, describes, explains, and understands reality. Modern science and technology disrupt the world they seek to know. It intervenes in nature, recoding it, capitalizing it, over-economizing it, and turning it into a useful resource for production and economic growth. The intervention of knowledge on the living planet marks and signifies reality with the inscriptions of metaphysical rationality, which constitutes a biotechno-power strategy set in motion by the domain of scientific-technical rationality in collusion with corporate capital. The illusion of illuminating the world, of representing it faithfully and accurately, of molding it and arranging it mathematically until it reaches the "truth," has simultaneously reproduced a dislocated reality that denatures nature, artificially transforming it through transgenic crops, animal clones, and the biotechnology of the food industry.

The scientific-technical activity in collusion with the financial leviathan empties the world and reduces it to a system of calculation and representation (Janke, 1988). It subjugates the land and turns it into a warehouse of assets, into a manageable and manipulable object, into a deposit of raw materials available for economic exploitation. The metaphysics consummated in modernity, expressed clearly in the Green Revolution, has transformed domestic animals into production machines and has turned the prodigal land into a pantry of natural resources, into a commodity devoid of poetic significance (Giraldo, 2013). Yet the agro-industrial empire is not content with exploiting the land for its use in accordance with its metaphysical rationality. It also aims to amalgamate the many ways people exist in a homogeneous model based on monocultures with genetically modified seeds, along with intensive use of pesticides and chemical fertilizers. Agroextractive globalization is an unhinged train that runs loose throughout the world, giving rise to huge waste and profuse pollutants that spread through the subsoil, water, and atmosphere.

However, "not even immense suffering which surrounds the earth is able to waken a sudden transformation," as Heidegger (1994c, p. 88) said. Western apprehension of the world is so strongly tied to the metaphysical way of understanding and experiencing everything that is not human, that desertification does not produce affections in the bodies of those who instrumentally plunder the terrestrial substratum. The destructive relationship, far from ceasing, uses increasingly sophisticated scientific-technical methods to continue undermining the last bastions of elements useful for economic valorization. Instead of feeling anguish, Western culture seemingly enjoys desiccation, natural extinction, and scorching of the earth. For metaphysical reasons, this devastation is not noted with sadness, because the earth is insensitive, a simple inert object; it is simply a resource lacking in the highest gifts that have been attributed to the lord of all other species, which, in spite of human folly, still inhabit the planet. Western culture, estranged from nature, values other living beings according to their capacity to satisfy the needs of a society sickened by the comforts provided by technological knowledge and economic rationality. The desolation that arises when nature is despoiled amounts to nothing more than a consequence of a way of life that obliterates our capacity to see ourselves as children of the earth, as having built a civilization detached from the ecosystemic fabric, of having undertaken a suicidal journey toward the light of reason in order to dominate nature and subject it to the whims of a few.

Land has been abandoned; it has become uninhabited. The landscapes of agroextractive business are monstrosities generated by a metaphysical reason that finds its genesis in humans' disregard for the fact that they belong to the earth. These landscapes are quadrangular agro-industrial farms that arise from the denial of the inhabited house. They are uniform, continuous, and inexpressive spaces that emerge from a way of being in the world that ignores the feeling of being on Earth. We human beings, blinded by our haste to subdue nature, are moving further and further away from considering Earth as our dwelling place (Pineda, 2016). We have instead created our own nature, a designed, simplified, disciplined, planned, and biotechnological nature. A uniform, legible, Fordist, and managed technology (Escobar, 1999) expressed in a paradigmatic way in the green deserts planted by agribusiness, where human beings have intervened in ecosystems to battle against nature, lifting reason to ever loftier heights while moving further and further away from the land. Agroextractive activity, whose reasoning perceives nature as a stock of dead resources that enter the productive cycle as inputs and justifies converting croplands into industrial factories, is an integral part of the effort to depopulate that characterizes the uprooting and alienation of contemporary human beings. The effects of the insatiable search to conquer nature as the only horizon of meaning are expressed in the body and landscape of a geometrically ordered agricultural surface, where imprints and scars appear on the skin of the earth left by agroextractivism, while humans exploit her but yet we refuse to inhabit her (Pineda, 2016).

Instrumental reason creates an artificial nature that can be calculated, measured, specified, and predicted so that, in the end, "commodities"are released as a presence that is unhidden by a technological package that ravages and consumes the earth. Thus, metaphysics is also expressed in its particular way of conceiving temporality: a particular way of looking to the future that guides the modernizing and agroextractive meanings of the modern world.

Progress and Development: Temporary Certainties of Agro-Capitalism

With Plato's philosophy, the road is cleared to exit the cave, reject the grounding to the earth, and define human destiny based on the world of ideas, and the meaning of one's sojourn in the "here and now" as a simple journey subordinated to the eternal world, which is where true human happiness resides. Such a metaphysical way of understanding reality—hybridized centuries later with Judeo-Christian beliefs—undergirds the West's attitude regarding its civilizational impetus during modernity for planetary domination with the help of modern science and technology. Nonetheless, in terms of temporality, it is worth referring to Aristotle's thought, since he demonstrates in his *Physics* how temporality, on which the notion of progress that defines modern culture is based, should be lived.

For Aristotle, time can be measured by the movement of a body according to the difference between the point of departure and the point of arrival. "The mobile leaves from here to get there," so that "the change takes place from this state to end up in that other state," writes François Jullien (2005, pp. 64–65). The key is understanding that the trajectory of the mobile that is measured between two moments represents the linear and irreversible temporality that characterizes Western culture. According to this particular way of giving meaning to the world, time can be understood by assigning two extremities, a before and an after, a beginning and an end, and a starting point and a finishing point, a path that is traversed linearly during a period of change and movement. Aristotle teaches that time is just a simple interval "where" one passes from one extreme to the other. However, the most important aspect of this temporal conception is not its starting point but rather the ending point, since the latter determines the meaning and direction of the whole journey. Arrival is the final goal. It is always the second extreme point that guides the path that one must follow from the first (Jullien, 2005).

We should not lose sight of the fact that Western culture draws not only from the Hellenic world but also from Judeo-Christian tradition, which has represented time in a manner very similar to the linearity of the Aristotelian mobile. For the Semitic people, time advanced from a starting point that could be traced back to the 7 days of the divine creation, ending at a glorious final point according to the plan drawn up by Providence (Nisbet, 1981). In other words, Christianity's belief that human destiny evolves like a preset plan, the meaning of which follows a journey that is taken according to a predetermined goal, i.e., resurrection of the dead, final judgment, and eternal life. This linear representation of time postulates a final goal of history, which determines the direction and meaning "through which" one journeys. Hence the origin of the meanings of "progress," "development," "goal with meaning," "ultimate end" (Janke, 1988), and all those teleological beliefs that we modern people share, according to which history is made by human beings so that we can evolve, like Aristotle's mobile, from a starting point incarnated in barbaric and primitive societies to ever-increasing civilizing stages.

This optimistic idea of civilizing progress is, without a doubt, the most insidious and persistent idea of modernity, which simply put is the belief that human history can be explained as a progressive development that goes "from less to more" (del Moral, 2004). It entails a certainty derived from the images of Christian paradise, where time runs from back to front, so that a secure meaning can be given to a human future, insofar as societies are moving in a civilizing journey in which all future time will be superior to the present. The notion of progress is a convoluted fusion of Christian beliefs of linear time with the profound reverence professed to science and technology since the European Enlightenment. Somehow, progress is said to be based on the link between secularized Judeo-Christian meanings and the advent of a new faith: faith in scientific and technical knowledge as the most effective means of improving living conditions of human beings (Sbert, 1996).

Modernity, in itself, is defined as an era in which there is a belief in constant improvement. Each novelty ages and is immediately replaced by a newer novelty, in a ceaseless, endless movement (Vattimo, 1998, p. 146). As Sbert (1996, p. 304) says, it is a "race in which even when one advances, the goal moves away," because progress is a linear and infinite path, which differs from the Aristotelian mobile, in that the final point is moved away just as one advances toward it. The present is a simple transition to the horizon, a mere fleeting moment of passage in the human odyssey on its way to the future. It is precisely here, in this peculiar way of understanding time, as if the present were coming from the past and moving to the future, where a fundamental aspect to understanding the logic of contemporary agroextractivism lies. Unlike cyclical time, in which one can always return to the origin, modern progress makes it impossible to return to the starting point, because the time line has been drawn in only one direction, i.e., away from the starting point, and no backtracking is allowed.

The problem is that this irreversible temporality has fatal consequences for human habitation on Earth, because the wandering nomad, traveling in the unidirectionality of the arrow of time, has to pack everything possible, given the need to foresee a journey without return. Thus, the extractive agribusiness entrepreneur must discount the present as much as possible and project it into the future. According to the metaphysical rationality of a journey that has no return to its starting point, it makes good sense to remove, undermine, and plunder all life and "resources" that currently lie buried, to accumulate and transport them to a time that has not yet come. For the linear time traveler who has embarked on a one-way journey, the best agreements dismiss caring for the underpinnings of the productive system. A good example is the land leases made by the Argentine soybean harvest pools: these contracts apply the reassuring formula, "extract, accumulate, and leave," in a manner similar to our time pilgrim's one-way journey.

Extractivism is based on these temporal meanings, since any exploitation of the earth can be explained in terms of a future paradise. For those who have constructed their symbols based on the notion of progress, there is an accelerating movement toward "development," so undoubtedly some environmental impacts are self-justifying for the sake of a more economically prosperous future. If history continues in permanent evolution, according to the believers in progress, the "advances" of modern technologies will be able to remedy unavoidable evils for the accumulation of capital and the economic well-being of future generations. In future-oriented temporariness, present sacrifice and the accumulation of useful "resources" take on significance because in the future one will be able to live better than in the past, thanks to the imaginaries promoted by development and scientific progress (Giraldo, 2014). Thus, the burdens of tradition will inevitably need to be overcome, because, for the devotee of progress, past and present are constantly renewed, bringing new eras, as nineteenth-century optimism dictates.

For the religions of progress, the past is an inescapable burden that must be overcome. According to their teleology, the barbaric, backward, savage, and underdeveloped societies that inhabit Third World countries in the end will have to modernize and succumb to the "champions of reason and progress" (Escobar, 2007, p. 98). All beliefs unrelated to the knowledge system of Western modernity are doomed to disappear. According to their peculiar way of living temporality, the predominance of scientific-technical reason, i.e., the basis for the optimism placed in progress, will sooner or later replace all those forms of knowledge that are not compatible with economic rationality (Escobar, 2007). Thus, the indigenous and peasant populations will have to adopt the metaphysical values of Western knowledge, or risk being defined as underdeveloped, backward, or "in the process of development", as established by a preset path along which all peoples must advance. For the future-oriented linear temporality, the peasant condition of the settlers of the Global South can be explained as a transitional state that will be inexorably overcome, as the history of other nations that long ago traveled this same route shows (Sutcliffe, 1995).

Modernization will thus be the only force capable of destroying superstitions and eradicating archaic relationships. All that remains is for traditional farmers to be provided with capital, infrastructure, and technical assistance to increase their productivity. So the progress of their economies and the maximization of their yields can be achieved by bringing technology into the equation, by means of, for example, the "extension" of Green Revolution packages. With these discursive certainties, the machinery of rural development began in the second postwar period, continuing the notion of enlightened progress, but oriented toward the countries of the "Third World." According to the World Bank (1975, p. 3 cited by Escobar, 2007, p. 275), this could be defined as a strategy "concerned with the modernization and monetization of rural society, entailing the transition from traditional isolation to integration with the national economy." Thus, as we will detail in the next chapter, from its origins rural development has involved a rationality of growth, an injection of capital, and the transfer of the nascent Green Revolution technology, in order to expand agroextractivism and the emergence of a remarkably uniform industrialized food system. The goal is for the "backward" populations of the "advanced" West to emerge from their lethargic past—as if they were children in need of adult guidance. This notion caused peoples of the so-called Third World to see themselves as inferior, underdeveloped, and ignorant and to doubt the value of their own knowledge and cultures (Escobar, 2007).

As Boaventura de Sousa Santos (2009) says, the belief in linear time promotes the idea of development, and the view that the peasant who farms the land is in a backward state and in need of help. It views time as coming out of the past and moving irreversibly toward the future. This understanding of the world is confident that history has the meaning and direction formulated by the scientific-technical knowledge of metaphysics. Moreover, "the central countries of the world system and, together with them, the knowledge, institutions, and forms of sociality that dominate them" are at the forefront of the race toward civilization (Santos, 2009, p. 110). This future-oriented temporal rationality views as backward all that is not congruent with what it has declared as advanced. Not only is knowledge from other cultures marginalized and disqualified, temporal rationality imposes the perception that its knowledge—a metaphysical knowledge as we have seen—has been formulated in terms of progress and development. In the case of agriculture, this means the imposition of a package of recipes from the petrochemical, pharmaceutical, biotechnological, and mechanical industries. The forms of existence that are not coherent with this technological and economic tyranny are considered unproductive, inferior, ignorant, and true "obstacles with respect to the realities that count as important: the scientific, advanced, superior, global and productive ones" (Santos, 2009, p. 112).

The superiority assigned to anthropocentric and metaphysical scientific-technical knowledge of European origin and the exclusion, omission, and silencing of subaltern knowledge, including, of course, indigenous and peasant knowledge prior to the era of rural development, are key aspects of the power relations of agroextractivism on a global scale. With the epistemic colonization by the center of its periphery, one might be swayed into thinking that the cognitive, technological, and social systems of the West are in a more "advanced" stage than non-Western knowledge and thus epistemic obstacles to be overcome, a view that is often shared by those in a subaltern position. In the end, the space of legitimacy of knowledge ends up being completely occupied by the knowledge generated by a scientific elite that considers itself to represent the world in the only true and valid way, by trying to place itself on a neutral observation platform (Castro-Gómez & Grosfoguel, 2007). In short, the industrial agribusiness that expands its tentacles over the world's croplands cannot be thought of exclusively as a technological and economic-political system, but as a complex relationship of cultural meanings that sustains those same structural configurations that make it possible. In other words, the epistemic dimension, which has been briefly described and challenged during the chapter, is deeply associated with the political economy of agroextractivism. For this reason, social movements' struggles to wrest power from agro-capitalism would be incomplete if a struggle were waged solely in the sociopolitical field without considering the metaphysical meanings of capitalist modernity. As an alternative to the hecatomb caused by industrial agribusiness and the empire of the global food system (van der Ploeg,

2010), agroecology must be, above all, an effort to change that civilizing sense that Western metaphysics seeks to impose on the Earth, devouring all forms of existence and spreading unflinchingly to all corners of the planet.

We will continue to make this fundamental critique, analyzing more precisely how metaphysics and the economic rationality that underlies agroextractivism are functioning in coordination, as well as the ways in which this economistic knowledge builds truth about agriculture. That is the next step, as we continue to excavate the cultural foundations on which agroextractive business is built.

Chapter 2
The Economic Rationality of Agroextractivism

> And why does it look so sad?
> —It's the times, sir.
> Juan Rulfo

Our discussion began by examining metaphysical thinking, since it undergirds the meanings on which a civilization was build that is currently collapsing and disintegrating. The agroextractive model is a fundamental part of the crisis of civilization, for which reason our critique of the model cannot go to its roots unless we first deconstruct and reconstruct the modern rationality on which unquestioned truths are based. That is why we have begun to reflect on (1) how nature has been "unhidden" in order to serve global productivism; (2) the dissociation of Western culture from the immanence of the world, from which the modern belief of reason is derived as an instrumental means to dominate all that is human and nonhuman; and (3) faith in progress guided by trust in the scientific-technical intervention of every corner of the known realm and all that still remains to be known.

Agroextractivism seen as a constituent component of the history of metaphysics may well be characterized by the above beliefs, which, needless to say, are specific to Western culture and, therefore, cannot be generally attributed to the history of agriculture of all peoples of the world. In a very concrete way, we are criticizing modernity as expressed by the convictions, discourses, and practices of industrial agribusiness and the agribusiness system on a global scale. In reality, we aim our criticism at a specific way in which Western culture conceives of reality and posit an estrangement vis-a-vis how the human group that belongs to that civilizing matrix feels, lives, and is significantly located in the world around it. This is what we call "rationality" for lack of a better term, although we should note that it is not used here as a synonym for reason. Rather, we understand it as a notion that is firmly tied to the way one affects reality and is, in turn, affected by it.

O. F. Giraldo, *Political Ecology of Agriculture*,
https://doi.org/10.1007/978-3-030-11824-2_2

This is how we address modern rationality, from which a type of economic thought is derived, spreads throughout the world, and expands through the practices of every field of knowledge, including, of course, the political, techno-scientific, and socio-environmental relationships that are woven into agriculture. Therefore, this chapter will explain the intrinsic relationship between metaphysical thinking and "economics" and the way economic rationality manifests itself in the discourses and practices of the capitalist agricultural system. We will review a brief history of agroextractive expansion, food regimes, and the way in which development discourse helped establish a sophisticated colonial construct on the periphery, building truth about agriculture through notions of poverty, hunger, productivity, and agrobiotechnology. Later, we will discuss how turning nature into a commodity contradicts the same cumulative logic of industrial agribusiness and the way in which agro-capitalism tries to protect and restructure itself in the current environmental crisis. The objective is to make visible the way in which agriculture is capitalized, governmentalized, regulated, administered, planned, and made the object of expert knowledge by international organizations, the state, and agro-corporatism.

Economic Rationality and Agriculture

Economic rationality is certainly the most radical manifestation of metaphysical thinking and the most important reference point for discussing how Western culture understands and intrudes in all aspects of the modern world (Leff, 2004). By economic rationality we understand subjecting everything to the laws of the market, whereby human beings act in accordance to the profit motive. Economic rationality assumes that all members of society are organized in the interest of profit and that their actions can be understood by the production and distribution of goods while perceiving nature as a commodity to be traded in terms of market values.

Despite what one might think, this way of understanding reality is very recent. In fact, the economistic vision—which today seeks to permeate every corner of the world, even to the point of enclosing the nucleotide spiral to commodify life itself—did not exist in Europe or other areas of the world before the eighteenth century (Escobar, 1996). Before the emergence of the market economy—which founded economic rationality as we know it today—all societies had markets with some sort of economic activity. However, none of these economies had been market dependent and were far from being strictly economic. Although it is true that the institution of the market dates from the end of the Stone Age, its role in human life had always been secondary. Although the economy existed in these societies, they were by no means determined by it. Actually, there was no institution that was wholly separated from everything else, such as the field of "economics," nor was there any profit motive. On the contrary, the economic system was integrated into the social system and was governed by principles other than personal gain[1] (Polanyi, 1975).

[1] In European feudal societies, there was no individual interest in owning material goods nor was

According to Karl Polanyi, the transformation that would become today's pervasive economic rationality began at the end of the eighteenth century, when the regulated market was replaced by a self-regulated market. The free-market system required an institutional division, whereby the economic order would operate independently of the social sphere. This distinction marks a fundamental shift away from previous economic systems, since the economic sphere had until then been embedded in the social sphere. Thus, until the late eighteenth century, in these societies no autonomous economic institutions were separate from society or operated according to the principle of profit. That is why the nascent liberal system of nineteenth-century English society, characterized by the isolation of economic activity, marked a spectacular break from previous societies. Polanyi explains how this change involved creating a new discourse after which personal motivations would be guided by profit. All human behavior would be judged as if mediated by monetary transactions, and all relations would involve the purchase and sale of goods. The economic rationality that accompanied economic liberalism is based on the idea that people are simple suppliers and consumers who participate freely in markets that work best when there is no external interference. In this way, all individuals operate as traders, offering products, services, or their own labor, such that their income, and thus welfare, depends on agreements that occur in the market. This is the concept of a society governed by market forces wherein people's lives are ultimately determined. Instead of the economy being part of social relations, as had been the case in all previous societies, social relations are subordinated to the economic system. This is the foundation on which capitalism was consolidated during the nineteenth century, i.e., the entire society becomes an extension of the market, whereby human and nonhuman life is transformed into a simple commodity.

The economy thus becomes an autonomous institution that organizes the entire society, subordinating all aspects to rules, to the point of treating everything as a commodity (Bartra, 2008). This of course includes nature and work, which, strictly speaking, are not goods produced for sale on the market. Yet, in the self-regulated economy, both work, indistinguishable from human life, and nature, solely produced by itself, suddenly became commodities bought and sold on the market and regulated by market laws (Polanyi, 1975). In the words of Enrique Leff (2004), the economy, now a field operating independently of society, governs human life, having previously established a rationality that dominates the natural order of things in the world. This occurs through a symbolic strategy that aims to subject all orders of "being" to a code of economic value as a precondition for the productive appropriation of capital.

The "economy" is the result of the professionalization of this knowledge policy that displaces social problems to an apparently neutral field such as economic science. Smith and Ricardo inaugurated this science at the end of the eighteenth cen-

there any profit motive. Social action was motivated by the need to protect one's social position, and the possession of material goods was valued solely for this purpose. Anthropologists who have studied indigenous societies in depth agree that, rather than profit, the guiding principles of indigenous economic systems are reciprocity, redistribution, and domestic administration, that is, production for the benefit of a closed group. For a detailed discussion, see Polanyi (1975).

tury through a theoretical study of the most convenient allocation of scarce resources and the balance of production factors. The objective of classical economic science was to understand how to maintain growth of output, recommending capital accumulation, division of labor, technological progress, and trade (Escobar, 2007). Beyond explaining the economy's inner workings, our analysis seeks to understand how it governs social organization through the production of knowledge that objectifies both the human and the ecosystem dimensions through the commodification of work and Mother Earth.

The main characteristic of modernity is the appropriation of nature as if it were a resource for us to use according to our whims. With the emergence of economic science, nature continued to be subjugated to scientific and technical knowledge, but paradoxically it was banished from the economic system. Nature was seen as a free and never-ending good from which the economic system could draw its "resources" continuously, using them as mere inputs in the production process (Leff, 2004). Concurrently, Earth was transfigured into a sinkhole for the waste generated by the economic process (O'Connor, 2001). Classical economists moved away from the physiocrats who had defended agriculture as the only true source of wealth, deciding instead that labor had the unique capacity to build value. According to their anthropocentric signifiers, they ignored the fact that human society cannot create nature[2] through work and regarded the ecosystem as solely an infinite and free provider of resources for capital accumulation (Porto-Gonçalves, 2006). The economy as an independent field of culture, politics, or ethics was also detached from the material basis on which it depends, "to remain suspended in the abstract circuit of values and market prices" (Leff, 2004, p. 183).

Economic science created the scientific basis for representing the world as an orderly image, so that growth in production could be measured and expressed mathematically. This science became the way of signifying and giving meaning to all existence, even permeating its rationality in fields as dissimilar as biology, sociology, law, politics, engineering, and agrarian sciences. In reality, economic rationality penetrated not only the other scientific disciplines with its terminology and its particular way of conceiving human inhabitation, but, above all, it profoundly infiltrated the cultural assumptions that underlie the way we behave and perceive ourselves. The term *homo economicus* sums up well the instrumental apprehension, by which we consider ourselves as agents who act according to the most rational way of obtaining profits in the market and operate according to market laws. This modern way of thinking and living implies viewing our connection with everything nonhuman according to a symbolic relationship that separates the domain of society from the land that serves as its sustenance.

[2]As Carlos Walter Porto-Gonçalves (2006, p. 55) asserts: "No society produces oil, iron, lead, water, and other minerals, just as man does not produce days or nights, solar radiation, without which we cannot live. We are as a species largely extractors of oil, iron, lead, water, etc., but not their producers. To say that we are producers means that the existence of what is produced depends on our creative capacity. To say that we are extractors indicates that we extract something that we do not make, which implies using it prudently."

Hegemonic economic science is the favorite child of metaphysical thought. Given its immense capacity to expand throughout the world, it can easily justify administering the hypertrophy of output that ignores the immanence of life, while managing a belief system, based on the linear notion of time, which posits that economic growth without limits is the key enables people to enjoy material prosperity.

Agriculture, of course, has been subjected to this forced infiltration. The logic of economic rationality penetrates an agriculture that has existed for millennia, capitalizing it, objectifying it, and transforming it into just another commodity (Leff, 2014). Thus, food becomes a commodity traded on commodity markets and subject to market laws, as are all "sectors of the global economy." Once immersed in this reasoning, one no longer questions the value of a healthy and equitable production and diet, in accordance with the cultural characteristics of the peoples and the ecosystemic conditions of their territories. What matters is the production of raw materials for industry in order to obtain economic benefit, because the economic rationality of capitalism is mainly interested in money entering the dynamics of the productive process to increase its value as much as possible. Capital seeks possibilities where it can obtain the most substantial profits. Whether to invest in agriculture or in something else depends on the profit return to capital.

This is a system oriented exclusively to favor profit, displacing the value of use. Thus, even if used as food, maize, wheat, or milk will not be produced if their production is not competitive (Hinkelammert, 2002). This is a social system governed by the laws of the market, where the economy rules and society must submit to its dictates. It matters little whether expansive monocultures overwhelm ecosystems and cultures, because when agriculture is transmuted into merchandise, it changes its appearance to become a powerful machine designed to generate profit and not support human existence (Bartra, 2008). The very meaning of agriculture is corrupted: It becomes a powerful locomotive steaming ahead, shunting from its path everything that hinders its compulsive greed.

This economic rationality, which accompanies capitalist development, rapidly spreads throughout the world following the consolidation of the industrial revolution in the second half of the nineteenth century. Four centuries ago, however, nature had already been colonized with the dualistic thinking that initiated the era of capitalism on a global scale. Through the colonization of the European peripheries, agroecological systems were restructured according to the demand of the hegemonic centers. From the fifteenth to the twentieth centuries, all territories colonized by Europe were subordinated to the interests of the colonizing empires (Alimonda, 2011; Quijano, 2000). Thus, entire ecosystems were devastated in order to substitute monoculture crops, including sugar cane, coffee, cocoa, cotton, tobacco, or extensive livestock farming. What had hitherto been biodiverse ecosystems or complex agroecosystems—such as those found in Andean or Mesoamerican cultures—were devastated by encroaching large estates that operated as haciendas or plantations. The latter depended on the labor force of native communities and African slaves. Through a policy of organizing colonized territories, agricultural was reoriented to satisfy overseas European demand.

The colonial process of favoring a single plant or animal species, solely because of its strong demand elsewhere (Worster, 2008), began in the sixteenth century with the colonization of the American continent and the plundering of slave labor in Africa (Quijano, 2000). At the close of the fifteenth century in England, the dynamic center where capitalism began, direct dispossession was carried out by the enclosure of common pastures and the monopolization of land by a few landowners. Consequently, dispossessed peasants had to sell their labor force to the new landowners (Marx, 1946). As Luxemburgo (1967) suggested, capital accumulation goes through a phase in the same central countries where the social relationship between capitalists and wage earners is actively manifested. Yet, it also has a colonial phase abroad. In both situations, traditional agricultural systems characterized by the diversity of crops adapted to specific locations and megadiverse ecosystems—for example, Latin American tropical regions—were destroyed and replaced by monocrops and pastures, in a process of reorganization that, as Worster (2008, p. 72) says, "brought as sweeping and revolutionary a set of land-use changes as did the Neolithic revolution."

The way nature was transformed according to capitalist logic and metaphysical economic rationality focused on specialization, which evolved into a radical simplification of ecosystems. The birth and consolidation of agroextractivism are tied to market signals that led to an especially profitable single species being produced. This ties in with the belief that it is more rational to buy food on the market with the money that comes from the sale of the mono-harvest. The economic rationality oriented toward specialization is at the heart of capitalism, as Adam Smith rightly said, so it is not surprising that if this logic occurs in agriculture, then diversification is eventually lost and production of a single product is favored (Worster, 2008).

As the rationality of the emerging "economy" indicates, an agro-industry specialized in generating profits is born, whereby nature is reoriented to satisfy market demand. Although there had already been a geographical reconfiguration of agriculture in the colonized areas, liberal agro-extractionism emerged in the heart of English society at the end of the eighteenth century. At this time, a new wave of enclosures allowed new landowners to produce solely for the market, spurred by the food and raw material needs of the growing urban population. The expropriation of the commons and other land through dispossession forced displaced populations to make a living by selling their labor force in the cities, thereby integrating both land and human life into market rationality and its laws of supply and demand (Polanyi, 1975). Throughout the world, this abrupt process of labor and land commodification varied according to the historical characteristics and biocultural specificities of each territory.

In Latin America, for example, the expansion of the hacienda originated partly from the legacy of colonial latifundia but was mainly a consequence of nineteenth-century liberal capitalist expansion. During this period, landowners seized wasteland and usurped indigenous lands, establishing leasing and sharecropping arrangements similar to those found during the origins of English capitalism. Entire production systems were determined by the market, including the cattle-cereal farms of the Southern Cone of South America, plantations with ex-slave labor, business agro-industry with a salaried labor force, banana plantations, the wine companies of Chile, or the rubber companies in the Colombian Amazon.

We should also mention the impressive growth of agroextractive activity in the British colonies and the vast American plains beginning in 1870. The imperialist agrarian expansion that began in the sixteenth century accelerated dramatically during the final decades of the nineteenth century and the first three decades of the twentieth century, when entire ecosystems were eliminated by the expansion of agriculture and cattle ranching to provide food and raw materials for the growing urban market. This territorial expansion of colonial agro-capitalism, which was preceded by the imposition of white supremacy and the genocide of indigenous peoples, caused unprecedented soil erosion, tragically exemplified by the 1930s Dust Bowl[3] in the Midwestern United States (Holleman, 2016).

In short, beginning in the mid-nineteenth century, under the impulse of English liberalism, increasingly greater spaces were subjected to the modern capitalization of nature and commodification of the labor force. Here agro-industry played a vital role in supplying food for the emerging economic powers and providing raw materials for the growing expansion of capitalism.[4]

The previous relationship that bound people to their lands was displaced by metaphysical economic rationality and its knowledge, which favored arranging and disciplining nature through an ongoing process whereby agriculture specialized in producing goods for the market. This process—fundamentally unchanged since then—led to a drastic simplification of agroecosystems and an enormous loss of biodiversity. This transformation comes at the expense of the destruction of natural conditions that, ironically, the same dynamics of accumulation need in order to continue their incessant process of valorization.

Agricultural Overaccumulation and the Development Project

Philip McMichael (2015) identifies the impressive expansion of agroextractivism that began in 1870 as the first global food regime. For McMichael, the expression "food regime" refers to how capitalism depends on specific food circuits that support the expanded reproduction of capital and the exercise of particular forms of power. According to the author, the first free-trade food regime promoted by the British Empire involved establishing monoculture agriculture, highly specialized in its colonies, and exporting its colonial output to Europe to supply the emerging

[3] The Dust Bowl was an environmental disaster caused by erosion stemming from the agroextractive model and a drought that lasted from 1932 to 1939. Eroded soil was blown up by the wind in huge clouds of dust that hid the sun. The phenomenon worsened the social effects of the Great Depression throughout the American West, causing nearly three million farmers to leave their land (Worster, 1979).

[4] The socialist essays of the twentieth century were also deeply guided by economic rationality. In these models of state ownership of the means of production, the same features of market-driven agro-extraction and technological progress were reproduced to maximize productivity.

industrial classes. Since then, the aim has been to maintain cheap food for the growing waged labor force and ensure the legitimation of the imperial sociopolitical order.

This period of rapid growth, ending after the outbreak of the First World War, was marked by an expansion of colonialism, such that in 1914 the colonial powers and their colonies expanded to 85% of the Earth's surface (Holleman, 2016), a consequence of the colonial order that structured food production and consumption in most of the world. The war would change the global geopolitical framework, however, with the main outcomes being the displacement of Great Britain by the United States from world economic leadership, the triumph of the Bolshevik Revolution in Russia, and the end of economic liberalism that would degenerate into the rise of fascism.

The First World War drove a spectacular growth of agriculture in the United States that continued throughout most of the 1920s. Agricultural production grew even as European demand, which had increased during the war, contracted over the years. In this way, agro-capitalism was affected by an overproduction of agricultural surpluses without demand growing at the same rate, resulting in an abrupt fall in profit rates that lasted throughout the 1930s (Mandel, 1972). To address this crisis and as part of the New Deal measures, President Franklin Delano Roosevelt, who became the US president in 1933, subsidized the decline in production while ordering crops to be destroyed to raise prices and surpluses to be purchased. Yet the problem of US overproduction was only temporarily alleviated during the Second World War and the early postwar years, when the surpluses were first sent to US allies and then used for the reconstruction of the European continent.

However, the structural problem of agricultural oversupply that had plagued the United States since the mid-1920s was still unresolved. In fact, the oversupply caused a great deal of justifiable fear as reflected by the prolonged and acute depression of the previous decade. The problem was how to find appropriate, long-term markets to remedy the overaccumulation, both of agricultural products and inputs and the industrial supply of the ascendant US empire, without undertaking—now—unfashionable colonial expansion. Indeed, the expansionist rivalry of the colonial powers had triggered the two world wars of the first half of the twentieth century.

It is in this context that we see the deep meaning of the "Development of Underdevelopment" project launched by Roosevelt's successor, Harry Truman, during the latter's 1949 inaugural speech. In the speech, Truman asked how the most prosperous countries could help more than half of the world's population to overcome undernourishment and poverty and proposed a technology-led development program focused on increasing productivity. The purpose, as Escobar (2007, p. 20) mentions, was "replicating the world over the features that characterized the 'advanced' societies of the time—high levels of industrialization and urbanization, modernized agriculture, rapid growth of material production and living standards, and the widespread adoption of modern education and cultural values."

Overnight, poverty became the greatest scourge to be addressed by efficient US-led aid. The objective was clear: to lead poorer countries—henceforth called the

Third World—down the unequivocal path of progress. Never before, stresses Gustavo Esteva (2009), has a word like "underdevelopment" been so universally accepted from the first day it was uttered. Truman coined it to name a calamity, an undesirable condition, called poverty, affecting most of the world's inhabitants, which the more "advanced" nations could help to alleviate. The same discourse that diagnoses the disease, i.e., poverty, contains the formula for its cure, i.e., development. Thus, the strategy defines Third World countries according to the economic patterns of the First World in order to justify intervening in their affairs (Escobar, 2007). The peoples of the Global South were defined not by who they were but by a multiplicity of problems that were attributed and then legitimized by institutional intervention in order to help "develop" them. In Foucauldian terms, development discourse first established the abnormality and then established control mechanisms over the abnormalized. It dictated what was unacceptable—the lack of economic conditions, the absence of modern technology, and the lack of education—and then prescribed the remedy, insertion into the monetary economy, provision of universal services, modernization, and industrialization. Thus, it defined the poverty of a population as underconsumption and then guaranteed an increase in the demand for certain goods and services so that the world's population could enter the monetarized economy (Illich, 1996).

The intention of the development campaign was to create consumers in Asia, Africa, and Latin America in an effort to boost effective demand. Capital had failed to boost low effective demand so as to compensate the increase in production in the first half of the twentieth century. With the "war on poverty" as its flag, the aim was to repurpose the colonial war machine for a huge "peaceful" project of aid to the "underdeveloped" world, to transform it into a massive market for the goods and services of the Global North (Illich, 2006a), while containing the advance of communism. Programs such as the Alliance for Progress sought to increase consumption of the emerging middle class based on the dogma that increased consumption leads to personal happiness, that new is always better, and that the purpose of life is to accumulate ever more wealth (Fromm, 1978).

Therefore, over time the object of desire was stimulated, and the population built expectations of matching the lifestyle of the North American middle class. Development was sold as a promise. The goal "to be developed" was rapidly accepted by the peoples of the South as a legitimate aspiration. The scheme was a machine that produced desire and regulated life based on acquisition, where needs were derived from an absence of consumption while promoting a deep fear of privation (Deleuze & Guattari, 2004). Thus, a sophisticated system was created that first stimulates people's expectations and then offers to fulfill their desires through multiple institutional interventions in the—abstract—name of development (Illich, 1996).

The poverty-representation regime was a prerequisite for creating the image of peoples who were undercapitalized, lacking essential goods and services, unable to manage on their own, and in need of help from those leading the race for progress. If the goal was to stimulate demand for the overproduction of goods, it was necessary for people to achieve certain consumption objectives. Needless to say, development was an unfulfilled promise, the wishes of the majority were not fulfilled, and, at most, what emerged was a modernization of poverty, insofar as people were consequently unable to live outside the market economy (Esteva, 2009).

The principles that structured the postwar world geopolitical order consisted of calling poverty an absence of development and affirming that the countries of the North would be the champions of civilization on a global scale. Perhaps the image of hungry people is the clearest expression of this sophisticated colonization mechanism. The photographs seen around the world of starving children with their extended abdomens, surrounded by flies, are exemplary of the image promoted of a Third World in need of food-aid packages that were created in the 1950s. This depiction hid the real intention of aid, which was to ship abroad the saturated supply of cereals that flooded the United States, as the FAO report *The State of Food and Agriculture* made explicit in 1954.[5]

To address this situation, the United States established policies for the placement of agricultural surpluses abroad at the rate of $650 million annually (FAO, 1954), and, in 1961, the United Nations created the World Food Program (WFP) with the same purpose. The maxim "feed the world" became the slogan that summarized the new program based on development discourse, through which countries in Asia, the Middle East, and Latin America increased their imports of US wheat, from 19% in the late 1950s to 66% at the end of the 1960s (McMichael, 2015). Thus, under the pretext of helping poor countries, most nations subjected to this plan ceased to be food sovereign, becoming net food importers (Ng & Ataman, 2008) and dependent on processed foods that standardized the world diet.

The strategy, in addition to providing cheap food to promote industrialization on the world's periphery and fulfill the economic goal of remedying overaccumulation, also had the political objective of controlling countries through food dependence and establishing an informal empire in the context of the Cold War. Perhaps President George W. Bush best summarized in 2001 the importance of maintaining a food empire:

> It's important for our Nation to be able to grow foodstuffs to feed our people. Can you imagine a country that was unable to grow enough food to feed the people? It would be a nation that would be subject to international pressure. It would be a nation at risk. And so when we're talking about American agriculture, we're really talking about a national security issue.

[5] "The feature of the food and agricultural situation which has caused most concern in 1953/54 has been the continuing accumulation of excess stocks of some commodities, particularly in North America" noted the FAO document (1954, pp. 28–30). "The surplus of cereals, particularly wheat, is much the most serious," the report indicated. "The problem was subsequently addressed by the FAO Committee on Commodity Problems, which discussed how the disposal of surpluses might be used to raise nutritional standards, e.g. of vulnerable groups in under-developed countries or to aid economic development, and defined the general principles to be observed in the disposal of agricultural surpluses… In the long run, the problem over time is how to avoid the recurrence of surpluses, once existing stocks have been liquidated, and how to achieve greater stability of markets. In other words, how production can continue expanding to meet the growing requirements of the ever increasing world population and the need to raise nutritional standards."

For the latter purpose, the *Global Campaign Against Hunger*, a successful fertilizer dissemination program, was established in 1963, and international agricultural research centers were set up to develop high-yield hybrid seeds. Advances in this regard were impressive: In just 20 years "almost half the land devoted to wheat and rice in countries of the South was planted with these varieties" (Holt-Giménez & Patel, 2009). This is how Green Revolution technologies and fossil fuel-based agriculture were made increasingly available throughout the world.

In 1973, in the context of the oil crisis, the United States and the Soviet Union signed a program called "Wheat for Oil," which meant less food was available for food assistance (Holt-Giménez & Patel, 2009). Perhaps moved to balance the macroeconomic effects of this shift brought on by the crisis following the hike in oil prices, World Bank President McNamara announced in 1973 a new Rural Development Program focused on increasing the output of small farmers by expanding Green Revolution technologies. The aim was to create a massive new group of consumers for agronomic and veterinary inputs of the Green Revolution and to turn them into market-oriented small entrepreneurs. According to World Bank development strategy, it was enough to modernize peasant practices through technical assistance, to specialize production toward commercial crops and animal breeds, to monetize rural society through credits and subsidies, and to integrate traditional economies into capital accumulation (Escobar, 2007). The size of such a project was not small: "The Assistance to the Rural Poor scheme of the 1970s claimed to help 700 million small farmers… with credits and green-revolution technologies" (McMichael, 2015, pp. 112–113).

Consequently, the Green Revolution invaded the world. At first with spectacular increases in production—food per capita rose by 11% between 1970 and 1990—then with more modest increases from 1983 to 1993 (Holt-Giménez & Patel, 2009), followed by a decline in areas where technology packages were first introduced (Ray et al., 2012; Pingali, Hossain, & Gerpacio, 1997). We cannot know for certain what the figure is, but some analysts believe that half of the world's farmers adopted the model. At this stage of rural development, nation states, advised by multilateral organizations, implemented policies that included increased research, agricultural assistance programs, access to agricultural credit, provision of rural infrastructure, and land reform to neutralize the advance of communism. These programs were implemented and continued until the end of the 1980s.

With the revival of the free-market model, agriculture was financed and converted into an object of speculation. The structural adjustments that were implemented intensively in the peripheral countries since the 1990s substantially reduced the political instruments of assistance. The model changed into social programs of direct monetary transfer, which put an end to self-subsistence crops and increased the consumption of a processed food diet bought with the money from subsidies.[6]

[6] By 2010, conditional transfer programs to the "poorest" had been implemented in 29 nations around the world. Also in 2010, in Latin America alone, they had reached 113 million people (Osorio, 2015).

In the United States and Europe, however, subsidies that exacerbated surpluses remained in place. With the new, multilateral, World Trade Organization (WTO) rules and the free-trade agreements, the tariff barriers of the countries of the South were eliminated, causing them to be flooded with cereals traded by agro-corporatism. Through dumping, subsidized output from the North ruined local farmers, since these products were sold at a price that was artificially lower than production costs, making it impossible for producers in the South to compete according to the rules that liberal capitalism had in theory established.

The development regime continued to function and govern overaccumulation. Yet since the rise of neoliberal globalization, the food regime, i.e., the political arm of capitalist growth, has been in the hands of transnational corporations, which have deepened the commodification of agriculture, expanding their operations to territories that hitherto had not been fully incorporated into world capitalism. This new global food regime was a continuation of previous systems, since, except for normative circumstances, conditions already generally existed for large corporations to step into the role that the state had played until then.

Through the lens of economic rationality that conceives of the world as potential output, economized geographies were established, and the lives of human beings were structured through hierarchical institutions that guided the expansion of capital from the mid-twentieth century on. This metaphysical process, linked to modernity, produced a landscape suitable for accumulation, but at the cost of destroying the conditions required for capital itself to maintain its tireless expansive dynamic.

The Environmental Consequences of the Commodification of Agriculture and the Restructuring of Agro-capitalism

> Do we still have to prove that our reason is doing violence to the world?
> Michel Serres

The liberal economy, detached from nature and designed to produce ever-greater profits, has no strict economic limits, insofar as the system always finds a way to resolve its own contradictions. The limits of an economy in which profits are its means and end, rather strange compared to pre-capitalist economic systems, and where the goal of investors is to obtain more profits, in a greedy, vicious circle of "money in search of more money," make up the natural and social conditions on which the entire economic process depends (O'Connor, 2001, p. 216). The commodification of nature and people's lives as a specific characteristic of capitalism deteriorates the reproduction of human beings' social fabric and well-being.

Despite the deep-rooted metaphysical notion that has sought to lead us astray from the nourishing land, the underlying problem is that it is impossible to separate the dangers that threaten the reproduction of life from the dangers that threaten the very existence of the human race. Although global warming, the destruction of biodiversity, geological restructuring, water pollution, loss of soil fertility, ocean acidification, deforestation, and the modification of biogeochemical flows (Rockström et al., 2009) will eventually generate a catastrophic scenario for humans and other life forms, previously these shifts will have destroyed the natural and social conditions required for expansion of the capitalist system. To commodify and reconfigure nature through technology in the image and likeness of industry implies undermining the social conditions of people's subsistence, as Marx foretold (1946, pp. 423–424) when he stated, "Capitalist production, therefore, develops technology, and the combining together of various processes into a social whole, only by sapping the original sources of all wealth: the soil and the laborer."

Agro-industry is an excellent example of how capital, as it manipulates nucleotides, standardizes crops and animal populations, mechanizes fields, and applies poisons and hormones, has not only generated atrocities; more importantly it has strangled production in the long term by exhausting the natural sustenance on which it depends while eroding the social base it requires for its reproduction.

To back these claims, we review some chilling data. According to the *Millennium Ecosystem Assessment* (MEA) report (MEA, 2005, p. 1), since 1950, "humans have changed ecosystems more rapidly and extensively than in any comparable period of time in human history." During this period, "More than two thirds of the area of two of the world's 14 major terrestrial biomes and more than half of the area of four other biomes" (Ibid., p. 4) were converted into monoculture land. Today, agriculture occupies 34% of the Earth's surface (Running, 2012). Although the World Wildlife Fund (WWF) says that the main cause of deforestation is cattle ranching and the expansion of agro-extraction, plantations of palm oil and soybeans head the list of the recent predatory actions. The WWF maintains that if this predatory trend continues, by 2050 some 230 million hectares of forest will have been destroyed, mainly in the Amazon, the Atlantic Forest and Gran Chaco, Borneo, the Brazilian Cerrado, the Chocó-Darién, the Congo Basin, East Africa, East Australia, the Greater Mekong, New Guinea, and Sumatra (Loh, 2000). This means 15 billion trees will be destroyed each year. If we continue at this pace, trees will have disappeared from the Earth's surface within 300 years (Crowther et al., 2015).

The agro-industrial model is also responsible for 70% of the world's water consumption (MEA, 2005), doubling in the last 50 years while increasing pollution of water sources (Running, 2012). Indeed, in some regions of South and East Asia, in the Near East, North Africa, and North and Central America, agribusiness uses more groundwater than can be naturally replenished. Such consumption is aberrant in itself but pales when we consider that water scarcity affects 40% of the world's population and that by 2050 the thirsty will include two-thirds of Earth's inhabitants (FAO-WWC, 2015). The acceleration of an economic system that is not based on the earth, but rather on the metaphysics of the laws of the market, violently increases water consumption, thus reducing its availability for plants, animals, and humans and degrading its quality to the point of making it inappropriate for the reproduction of life.

The agro-industrial system that expands without consideration for the world's croplands is dependent on mineral fertilizers, the production of which has steadily increased since the mid-twentieth century. Products used by agro-corporatism and by small and medium farmers who participate in the logic of the technological package have had disastrous consequences on the health of the ecosystem. The most widely used—and most harmful compound for ecosystem health—is synthetic nitrogen, said to reduce humus content, organic matter, and soil biodiversity. Nitrogen fertilizers are also responsible for the increase in soil acidity and the consequent inhibition of crop growth (Kotschi, 2013). Since soil health is the most important natural condition for agriculture, the decrease in its quality due to the extensive use of fertilizers is an inherent contradiction in the practice of agroextractive industries. The law of declining returns means that ever-increasing amounts of inputs must be applied for crops to reach previous output levels, as reflected by data that estimate that yields per kilogram of nitrogen have declined by one third since 1961, when chemical fertilizers began to be used throughout the world (Grain, 2009).

In economic terms, Ricardo and then Marx detected the contradiction whereby agro-capitalism, in its haste to exploit every element of the land, progressively reduces productivity by breaking the natural base required to ensure profitability. By treating nature as a commodity, capitalism exhausts the conditions necessary for its accumulation. This is why the loss of biodiversity, erosion, salinization, acidity, and compaction and, in general, the loss of soil fertility have brought a stagnation or decrease in yields, as producers have already experienced with the main cereals such as corn, rice, wheat, and soybeans.[7]

The increased use of synthetic fertilizers to replace soil degradation—which has increased 500% in the last 50 years (Running, 2012)—has not only degraded soil fertility, it has also polluted water, causing algae growth and the death of untold numbers of fish. Phosphorus, for example, is accumulating in agricultural soils, causing eutrophication in rivers, lakes, and ocean coasts (MEA, 2005). However, the worst effect of fertilizers on the environment is the emission of nitrous oxide: a gas involved in the destruction of stratospheric ozone and global warming. This gas has a greenhouse effect 300 times more powerful than carbon dioxide. In fact, for every 100 kg of nitrogen fertilizer used to fertilize the soil, 1 kg ends up in the atmosphere as nitrous oxide (Lin et al., 2011). The Intergovernmental Panel on Climate Change (IPCC) estimates that agroextractive activity is responsible for between 10 and 12% of global emissions that produce global warming. Estimates made by Grain (2013), however, indicate that if the total industrialized agri-food system is included, the figure could represent half of all emissions on a global scale.

Pesticides also threaten the very productivity of the system, as 75% of the most important crops depend on biotic pollinators, which are seriously affected by the increasing use of insecticides. There are an estimated 20,000 species of pollinators, many of which are threatened by the use of pesticides. The case of the dramatic decrease in bee populations in the United States and Europe and its close relationship with the use of pesticides and the planting of genetically modified crops is a

[7]A study that included worldwide agricultural censuses taken between 1961 and 2008 found that although yields increased in certain areas, between 24 and 39% of maize, rice, wheat, and soybean yields remained static or collapsed (Ray et al., 2012).

clear example of the self-destructive capacity of the system itself. Deforestation for the establishment of pastures and monocultures; privatization and contamination of water; chemical-synthesis fertilizers and their effects on the soil, atmosphere, and water sources; and the mortality on pollinators from pesticides are good examples of why treating nature as a commodity and converting it into a mirror image of the factory decrease the possibility of life reproducing itself. In its eagerness to conquer nature and tirelessly increase output, agricultural capitalism has not only degraded the health of the Earth; paradoxically it is also reducing the return on investment, which is a contradiction of the system itself, as it contravenes the reason for the existence of capital (Bartra, 2008).

This contradiction is what James O'Connor (2001), recalling Marx's aforementioned aphorism, calls the second contradiction of capital. For O'Connor, capitalism cannot refrain from degrading its conditions of production and putting its own profits at risk. In a self-destructive way, capital tends to affect its ability to produce and thus accumulate more capital. This has led to a decrease in returns because it is destroying the basis of its very existence. Capitalism is its own barrier, says O'Connor, because hyper-productivity ultimately does nothing but raise costs and lower the return on investment.

Interestingly, agro-capitalism's flexibility allows it to adapt to different scenarios. The system takes advantage of the crisis to try to restore its own production conditions. To do this, it makes numerous adaptations in order to lower production costs and increase profits. These changes go by various names: sustainable intensification, climate-smart agriculture, precision agriculture, drought-resistant transgenic production, and organic agriculture based on commercial inputs, and more recently, it has sought to co-opt agroecology stripped of its political content (Giraldo & Rosset, 2017). These many transformations fall under the geopolitics of sustainable development, whose objective is to try to protect capital from its worst excesses and open new sources of business, such as carbon-capture payments, agro-ecotourism, bio-trade in native seeds, organic monocultures for sale in large areas, and what in the future could be called "the agroecological input industry."

Agricultural capitalism, with the help of governments and multilateral organizations, is in a process of transformation aimed at solving the crisis in its benefit by taking technical steps to recover the degraded substrate. It is also establishing new planning instruments to lower the costs of labor force reproduction, including greater flexibility in agricultural labor markets, or the establishment of partnerships between small, medium, and large farmers, as we will see in the next chapter.

The environmental debacle is an excellent opportunity to legitimize a renewed dual agrarian structure under the mask of green discourse and socio-environmental resilience. It seeks, on the one hand, to confront the second contradiction by attempting to restore agro-capitalism's production conditions, while, on the other hand, to incorporate small farmers into the market by involving them in commercial chains, now presented as environmentally and socially responsible investments. As the abovementioned data show, the technologies of agricultural capitalism have fractured the ecosystemic order, requiring certain activities to be "green washed," thus legitimizing a false environmental discourse that appears friendly to peasant populations.

We are witnessing a new phase of the Green Revolution and a reconfiguration of the global agricultural model in the context of the contemporary environmental crisis. The changes induced by the crisis include technological arrangements that hybridize agricultural and veterinary biotechnologies and agroecology and co-opt proposals defended by people, combined with the classic state policies that facilitate the conditions of production. These include the provision of capital, credits and subsidies; deregulation of obstacles to access to land, legal changes aimed at eliminating obstacles to purchases, leases, or concessions of vacant land; provision of infrastructure, roads, irrigation districts, and port construction; incentives for capitalization, purchase of machinery and agricultural and veterinary technology; and investment in agricultural research, biotechnology and process innovation.

In the short cycle, as Braudel would say, these adjustments bring capitalism out of the crisis and restore its ability to accumulate. However, in the long cycle, this restructuring deepens devastation, since thinking oriented to the scientific and technical exploitation of the world continues masking the dominant assumptions about nature that have been inherited from modernity. The watchword involves finding new technology and planning tools to continue extracting the maximum wealth from the earth, while avoiding the environmental and social limits that capital has encountered.

Growing social dissatisfaction caused by an advancing agro-industrial sector (which has fractured metabolic relationship between human beings and the land) is placated by controlling the discourse of the people's struggle. By skillfully manipulating the message, the agro-industrial sector can often turn around its adversaries' demands so that they coincide with the dichotomous premises of sustainable development and economic rationality. This linguistic control can have the effect, as O'Connor observes, of making these movements quite functional to the system, since they can save capital from itself by forcing it to confront its negative effects, encouraging the accumulation model to become more efficient. Capitalism is not organized in response to a central decision-maker; among other factors, it is the result of several movements that rally against the laws of the market. For this reason, when opponents are mildly reformist, unable to think outside the "market box," they become allies of capital by helping to safeguard it from its worst contradictions.

Other movements, however, have understood that the struggle is both political and epistemic, and in this sense they have reconfigured the contents of their dissent against capital's desire to mold ecosystems according to the laws of the market. Food sovereignty, peasant and popular agroecology, the struggle for land and territory, and the defense of life and Mother Earth are just some heuristic concepts created by movements in different latitudes. As we shall see in Chap. 5, these issues are challenging capital's efforts to reorganize itself in light of the current environmental crisis.

Predatory agroextractivism, with its symbolic basis in metaphysical economic rationality, has crossed all boundaries. We find ourselves in a crisis that has exceeded all limits, which the hegemonic system itself recognizes and thus realizes that it does not have the luxury of rejecting any available tool. We therefore face a scenario in which the world agricultural system is being restructured based on incorporation, co-optation, and capture, through a process of regulation and administration on a global scale governed by a metaphysical economic rationality whose expansion will require detailed analysis.

Chapter 3
Territorial Control and Geographical Expansion of Agribusiness

> In this society, the productive apparatus tends to become totalitarian to the extent to which it determines not only the socially needed occupations, skills, and attitudes, but also individual needs and aspirations.
> Herbert Marcuse

In the previous chapter, we reviewed the conditions that allowed the spread of agro-capitalism over the land and the environmental and social effects of that expansion. We saw that the rationality of economic metaphysics is agro-capitalism's foundational substratum and how certainties were created by the governmentalization of agriculture on a global scale.[1] At the end of the last chapter, we addressed how the system feeds on the crises caused by its own self-destruction, and the mechanisms it uses to reinvent itself, as it finds new spaces to continue its process of accumulation.

In this section, we continue discussing the territorial growth of agro-capitalism but now looking more closely at the phenomenon of land dispossession in the fields of the Global South. In general, I show that, in spite of the most violent dispossessions at the dawn of the millennium, territorial control is more effective when it occurs by incorporating, not excluding, when it is based on the functionalization of

[1] By governmentality, we follow Foucault (2006), who coined the term to explain how government technologies do not reside in the state, understood as an actor formed by a set of institutions that support government. Foucault is interested in conceiving government as a multiplicity of practices endowed with a specific rationality. That is why, far from understanding neoliberalism as a "withdrawal" from the state, we will approach it, together with Foucault, as a "way of doing things," the meaning of which makes the state an instrument to create market autonomy. In other words, we will be dealing with the governmentality of neoliberal agriculture as an art that seeks to generate the conditions so that individuals can act freely, as long as this is done in a way that is consistent with the interests of economic accumulation (Castro-Gómez, 2010).

The key ideas in this chapter were published in Giraldo, O.F. (2015), "Agroextractivismo y acaparamiento de tierras en América Latina: Una lectura desde la ecología política," *Revista Mexicana de Sociología*, 77 (4): 637–662. This version widens the discussion and offers a global view of the phenomenon.

O. F. Giraldo, *Political Ecology of Agriculture*,
https://doi.org/10.1007/978-3-030-11824-2_3

the subjects with the consent of people who are hegemonized by the discourses and practices of agricultural development. Although violence is the most visible method, the most powerful forms of territorial control occur when they are mostly silent; when power is sought with the consent of the population as a whole. Through an imperceptible imposition of the technological, cultural, and representative regimes of agroextractivism, national states—guided by large institutions that largely control the world's agriculture—try to create the necessary conditions for the territorialization of the agro-industrial model.

We begin by discussing the problem posed by classical Marxism, known as the first contradiction of capital, and how this difficulty is addressed through dispossession. Subsequently, we discuss rural ontologies and epistemes through a phenomenological interpretation showing the nodal points where the separation of populations from their material and symbolic conditions of existence occurs. We then undertake a phenomenological treatment of the notion of territory, to examine the meaning of deterritorialization without physical displacement. Grounded by these theoretical constructs, we will interpret the narratives of international cooperation organizations on agricultural policies in the coming years, with special emphasis on the inclusion of the peasantry in the dynamics of neoliberal globalization. We will conclude by trying to give a satisfactory answer about why huge efforts are made to include small farmers around the world in the system using territorial rents. In short, our interpretive exercise will study how agro-extractionism uses different strategies to expand without directly appropriating all land.

Crisis of Capitalism and Violent Dispossession of the Land

Capital requires constant geographic expansion to escape its intrinsic tendency to enter recurrently into crisis, because its productive capacity tends to increase much faster than effective demand. As Marx explained in his theory of the fall in profit rates (1946, p. 213 ff.), capital is prone to crises because of its propensity to produce more than it can consume. These crises are manifested as an oversupply of goods that cannot be sold without losses in the domestic markets and in surpluses of money that do not have profitable investment opportunities (Harvey, 2007). As we explained in the previous chapter, this occurred during the great depression of the 1930s and was a reason behind the creation of a large postwar development project. The Development of Underdevelopment rests on the premise that the countries of the South need to be redeemed by the more advanced nations—especially the United States—through international aid. This discourse created a "peaceful" neocolonial strategy to transfer underproductive capital to countries designated as "underdeveloped" and create new consumers for products and services of the Global North.

Since the previous colonial strategy of occupying and controlling more territories was outdated—having provoked two world wars—the development crusade was an updated global maneuver to conquer new markets and utilize excess capital through hyper-urbanization and complete subordination of the countryside to the city. Without the constant geographical expansion of capitalism, its permanent extension toward new and diverse territories, and the incessant production of landscapes guided by economic rationality, capitalism is incapable of surviving.[2] Capital is designed to expand ceaselessly: a metastasis that never refrains from invading any territory of global geography. However, this expansion proceeds unevenly while organizing space with regional differentiations and specializations (Harvey, 2003). A center-periphery dialectic, in which the dominant centers subordinate the production of the space on their periphery, which is what occurred with the world agricultural reorganization following the conquest of America (Wallerstein, 2010). Even though this discourse conjures up an image of a race in successive stages where some countries are at the forefront of the race of modernization (Sutcliffe, 1995), the truth is that the difference between countries is not in the advantage they have in advancing toward the abstract goal of "development." Rather, it involves the role that each nation—or region within the same country—plays within the same system: supplying raw materials or specializing in higher value production (Beigel, 2006).

Despite the apparent functionality of this unequal organization in the midst of which globalized capitalism is growing, the model is not free from contradictions. The expansion of the system as a whole has created a pathological hypertrophy that is seriously threatening the health of the system, seen in the light of the falling profit rates of globalized capitalism. The data analyzed by Robert Brenner (1999) have shown that since 1973, capitalism has been suffering from a chronic crisis of overaccumulation. This has happened because the permanent increase of capital has a self-destructive effect on the system itself: international hyper-competition that generates a decrease in profitability rates. The entry of new industrial powers into the world market—first Germany and Japan, then Southeast Asia, and finally China—brought forth the same goods and services that were already being produced by the countries that first industrialized but much cheaper. "The result" argues Brenner (2009, p. 14) "was too much supply compared to demand in one industry after another, and this forced down prices and, in that way, profits." Of course, companies have embarked on a race of technological innovation, creating new processes and product variations—scheduled obsolescence, among others—and have moved geographically to the countries of the South in search of spatial-location advantages to lower costs or have increased mergers between firms to form oligopolies, but in the end, the rate of return has only worsened.

The easiest way to overcome a lack of attractive investments and the ensuing idleness of capital surpluses was through financialization. This allowed capital to bypass, as O'Connor (2001, p. 288) asserts, "the long and tedious process of leasing

[2] In the words of Marx and Engels (2001, pp. 25–26) in the *Manifesto of the Communist Party*: "The need of a constantly expanding market for its products chases the bourgeoisie over the entire surface of the globe. It must nestle everywhere, settle everywhere, establish connections everywhere…It compels all nations, on pain of extinction, to adopt the bourgeois mode of production; it compels them to introduce what it calls civilization into their midst, i.e., to become bourgeois themselves. In one word, it creates a world after its own image."

factory space, buying machinery and raw materials, renting land, finding the right kind of labor power, organizing and implementing production, and marketing commodities." Instead, financialization encouraged speculation in financial markets and electronic money. To escape from the "real economy" and from having to safeguard the international financial system was a good way to overcome the chronic problems of overaccumulation. However, the basic remedy was to implement a strategy of plundering, supported and promoted by national states through neoliberal privatization policies, in order to transfer public or common assets to private companies, and then insert them into the private flows of capital accumulation. This process re-creates on an enormous scale the common grassland enclosures of sixteenth-century England described by Marx under the label of "original accumulation"[3] and renamed by geographer David Harvey (2007) as "accumulation by dispossession," in order to stress that it is a process that continues today.

According to Harvey, the violent separation of people from their living conditions as described by Marx, such as the privatization of land and the expulsion of peasant populations; the transfer of people's common goods into private property rights; the suppression of access to communal goods; the commodification of the labor force; and the colonial appropriation of nature are not a foundational anecdote of capitalism, but rather ongoing processes that have deepened in our days. Harvey's argument can be summarized as follows: During the current global crisis of overaccumulation, capital owners do not find profitable sources of investment, so they must resort to another strategy of having assets released at a very low or null cost, in order to take them over and derive profits from them. This is none other than shameless looting and pillaging that seek to privatize everything that can be inserted into a value-generating circuit, without remunerating the people to whom everything belongs.

[3] Marx (1946, p. 510) claims that English serfdom had practically disappeared in the last part of the fourteenth century, which is why "the immense majority of the population consisted...of free peasant proprietors, whatever the feudal title under which their right of property was hidden." However, this situation would not last long, since the prelude to capitalism coincided with the desire of royalty to obtain absolute sovereignty, for which reason it had to accelerate the dissolution of the feudal armies. Faced with such a scenario, the lords rose up against the monarchy and parliament and violently threw the peasants off the lands they were cultivating, in addition to usurping the communal lands on which they grazed their cattle. According to Marx, another event that influenced the expropriation of the direct producer was the Reformation against the Catholic Church, since this institution owned a large part of English land. The Protestant onslaught caused many of the goods of the Catholic Church to be given away to a few individuals protected by the king and parliament or sold at a derisory price to land speculators. Either way, the result was that the villagers expelled from ecclesiastical lands and the usurped peasants were forced to work for others. In Marx's words: "The spoliation of the church's property, the fraudulent alienation of the state domains, the robbery of the common lands, the usurpation of feudal and clan property, and its transformation into modern private property under circumstances of reckless terrorism were just so many idyllic methods of primitive accumulation. They conquered the field for capitalistic agriculture, made the soil part and parcel of capital, and created for urban industries the necessary supply of a "free" and outlawed proletariat" (*Ibíd.*, p. 516).

Examples of this neoliberal "accumulation through dispossession" abound: enclosure of oil, mega-mining, and carbon and water; privatization of public health, pension, and education institutions; expropriation of people's knowledge to turn it into intellectual property rights; biopiracy and bioprospecting; and commercialization of cultural expressions. These activities exist for the accumulation of capital, starting with the dispossession of public and common goods that were institutionalized with structural reforms guided by multilateral organizations since the last decade of the twentieth century and intensified with the bursting of the financial bubble between 2007 and 2009. Speculative capital, which until then had taken refuge in the metaphysical financial markets, was forced to descend to earth and seek new sources of accumulation in nature. The extractive offensive of big capital involved particularly important investments in conventional and unconventional hydrocarbons, construction of dams, extraction of precious and industrial metals, and a subject that should be carefully analyzed: land dispossession in the countries of the Global South.

According to the LandMatrix platform (2017), from 2000 to 2017, an area of land that almost doubles the size of Ecuador changed hands. The huge investments recorded in 2008 were particularly important, when the rise in global food prices coincided with the collapse of the housing bubble and the global financial crisis. The scandal over land dispossession was so great that in 2009 the government of Madagascar fell from power due to a project presented by the Korean company Daewoo to acquire 1.3 million hectares on the African island. The phenomenon consisted of systematic land acquisitions in the peripheral countries of the world system by some governments—such as the oil-producing countries of the Gulf, China, India, Japan, and South Korea—pension funds, and illegal capital from tax havens in order to launder money and find new sources of speculation. The target of foreign hoarders has been Africa, Latin America, East Asia, and Oceania (Grain, 2016), in addition to the concentration of land by national elites in their own countries. The main use of this accumulated land has been the sowing of flexible crops, i.e., multifunctional crops used for human and animal consumption, industrial material, or agrofuels—such as palm oil, soybeans, and sugar cane—some of which vary in use depending on their price at the time of harvest (Borras, Hall, Scoones, White, & Wolford, 2011). Logically, the governments of the countries receiving the investments have been allies of big business and have facilitated the dispossession. Governments have made legislative adjustments, offered tax and customs incentives, modified institutions, kept wages low, and built infrastructure.

In addition to investments in agricultural and forestry monocultures, many hoarders may be buying up land in anticipation of a foreseeable rise in food prices in the coming years. Indeed, far from decreasing, international organizations that have investigated the phenomenon foresee that there will be growing interest in buying land for agricultural purposes. According to their estimates, the world's population will increase from the current 7.2 billion to 9 billion by 2050, by which time the global economy is expected to triple. This will lead to increase demand for food and agricultural products for nonfood use and, in general, a substantial increased demand to own and control the land. By land dispossession, we refer to the soil but also to water, subsoil, biodiversity, forests, and air. This constitutes an appropriation of the set of natural bodies on which we depend as biological beings, and which are indispensable to satisfy the growing demand for nature required to sustain the suicidal civilizing model in which we are immersed.

The fever to acquire land has accelerated ecological degradation. According to LandMatrix, one third of the land accumulated by investors was deforested, and according to Anseeuw et al. (2012), wetlands and mangroves are also subject to land conversion. In addition, hoarding has led to the displacement of entire communities, since approximately 45% of purchases were made to acquire small-scale farmland and livestock-grazing land (Oxfam, 2012), often carried out either through deceit[4] and threats[5] or private acquisitions encouraged by certain governments. Most rural communities in countries where dispossession has occurred lack land titles and often have customary land rights, making them highly vulnerable to eviction from inhabited land (Oxfam, 2011).

These maneuvers have led Marxist authors to conclude that capitalist relations of production, given the crisis that has deepened since the end of the 2000s, have made violent dispossession the main strategy of accumulation. Consequently, these authors have placed importance on revisiting Marx's explanation of original accumulation. This academic current, also incorporating Rosa Luxemburg's work, maintains that land-grabbing and expulsion of peoples from the Global South, among many other examples of dispossession, simply corroborates a repeat of the most violent, brutal, and cruel strategies of primitive plundering used to bring new areas of exploitation into the global circuits of capital accumulation.

I believe, however, that we should exercise caution regarding such claims. Although capitalism may occasionally appear to exercise impropriety and the most violent strategies of accumulation, I do not think we are on the verge of an end to the "peaceful" strategy inaugurated by President Truman, nor should we claim that capitalism has renounced its sophisticated strategies and made way for increasingly ominous methods. Although we cannot ignore the forced expulsion of many populations—especially in Africa but also Latin America[6]—through the most blatant form of original accumu-

---

[4] In Indonesia in the mid-1990s, PT. Menara Alpha Semesta negotiated with local communities to convert their land into palm-oil plantations. The agreement consisted of a 35-year lease, while the company built houses, schools, clinics, and sanitation infrastructure. Each family transferred 7.5 hectares to the company, five of which would remain in the company's hands, and two hectares would be returned planted with palm. Oxfam (2011) notes that the promised infrastructure was never built, only one hectare was returned to each family, and there is an agreement with the state that when the 35-year lease expires, it will be renewed so that dispossession will continue an additional 95 years.

[5] A saying that I often heard from the victims of paramilitarism in the Colombian Pacific and in Guatemala is, "Either you sell to me, or I will come back later and do business with your widow."

[6] An example of the violent land-grabbing offensive is the Polochic Valley in northeastern Guatemala. This territory comprises five municipalities, with 220,000 inhabitants, 89% of whom are indigenous Queqchí and Pocomchí. This fertile region provoked the interest of the sugar-mill

lation, I believe that capitalism has become increasingly subtle in its maneuvers, a tactic to which its proponents adhere. Today it is difficult for capitalists to achieve large-scale deals that involve huge exoduses, since media exposure makes them unsustainable. Forced displacement of rural communities and dispossession of land is such a difficult phenomenon to legitimize in today's world that it is very difficult for direct land expropriation strategies to repeat actions taken at the beginning of the twenty-first century. In fact, some of the big businesses involved in dispossession began to falter starting in 2008, and some studies have shown that forced displacements have become less common (Grain, 2016). Consequently, we should build awareness of the increasingly sophisticated mechanisms that modern capitalism is deploying to take over land and control nature.

The argument I want to make comes from Marx who posited that "original accumulation" is a historical process of decoupling the producer from the means of production, based initially on the expropriation of land from farmers, i.e., a violent fracture by which the peasant is separated from his land. Neo-Marxist authors have used this line of reasoning to highlight current examples of dispossession, and I will not dispute their argument. However, I would like to recall a comment that Marx made about his thesis, which is known as the Draft Response to the Letter to Vera Zasulich, in order to analyze the phenomenon in a slightly more complex way. In this text, Marx (1974) stated that while the secret of "original accumulation" was the radical separation between the producer and the means of production, it was also true that "in order to expropriate farmers, it is not necessary to drive them from their land." Although Marx here refers to another kind of violent coercion, in my opinion Marx is giving the key to understanding that expropriation depends on both physical expulsion and on dissociation that transcends direct dispossession.

In his letter, Marx wrote to the Russian Socialists that original accumulation would not necessarily follow the same sinister paths taken in England, because the vitality of Russian rural communities had proved incomparably superior to that of other societies. I will not pursue this rather culture-centered approach. However, in spite of the more orthodox peasant-less visions that have predicted the end of the peasantry, many peoples of the Global South—such as those that Marx praised in

owners and palm growers. Through purchases, leases, or agreements with the peasants, the agro-industrial companies acquired land, to the point that between 2005 and 2008, the Chawil Utz'aj sugar mill alone had purchased 5000 hectares of sugar cane. The communities in Polochic had no choice but to seek refuge in the Sierra de las Minas Mountains. In 2010, the mill had some problems with the loan it had acquired for the plantation, and the small farmers who had been evicted years earlier decided to return to their land. Yet in March 2011, private security units violently expelled more than 800 returned families and destroyed their crops, lives, and scarce belongings (Oxfam, 2011). Other emblematic cases of displacement caused by the territorial offensive in Latin America are the displacement by the Benetton company of the Santa Rosa Leleque community in the province of Chubut, Argentina; the violent dispossession of 30,000 hectares of land by paramilitaries in Colombia that belonged to the black communities of the Chocó Pacific region of Curvaradó and Jiguamiandó. There were displacements in other regions, such as Alto Mira in Tumaco and the south of Bolivar for the planting of African palm trees (WWF, 2009), or the dispossession by the Sunway palm company in the El Samán compound in the Province of Los Ríos, Ecuador (FAO, 2012a, 2012b).

nineteenth-century Russia—have shown their capacity to adapt and survive the onslaught of Western modernity, hegemonic capitalism, and the colonization of knowledge. Moreover, there really are quite a few. According to figures from the ETC group (2009), there are about 1.5 billion peasants, 190 million nomadic pastoralists, and more than 100 million fisherfolk, of whom at least 370 million are indigenous, who live on more than half of the world's arable land.

For capital, small producers' lands are an important source of accumulation, so it behooves us to analyze how and under what conditions Marx's assertion is true that in order to appropriate peasants' land, it is not always necessary to drive them off their plots. We have here a more profound way of interpreting land-grabbing in the world. This is not just a problem of direct dispossession—i.e., the most obvious and thoroughly described by contemporary Marxist theory—it is also as a way of splitting peasants from the land through the delocalization of their cultures and their subsequent re-functionalization, an objective achieved by multiple mechanisms that will be elucidated below.

Farmers in the Global South: Ontological and Epistemic Colonization

> The mode of human sense perception changes with humanity's entire mode of existence
> Walter Benjamin

To understand the mechanisms of dispossession in situ, it is worth clarifying that metaphysical rationality and the dualism between nature and culture cannot be blamed on all rural people of the Global South. This economic rationality, tied to a future-oriented linear temporality, the dissociation of the immanence of the earth, the trust in reason and technology, and the efforts to manipulate, subjugate, and dominate concern a very specific perspective of knowledge. Although such rationality has attempted to engulf other forms of knowledge and fit them into its structure of meanings, it has not yet succeeded in standardizing and matching the various ways in which people understand and explain their reality. The dichotomy between nature and society makes no sense to many local cultures in the Global South, for which the biophysical and cultural symbols as well as the supernatural are inseparably intertwined (Descola & Pálsson, 2001).

I do not want, however, to fall into the temptation of making cultures essential in the manner of the good green savage. I am referring to that invention—so widespread in many currents of the contemporary academic left—which holds that until the European invasion, indigenous peoples lived in balance and harmony with their natural environment. This is a colonial position that re-creates indigenous peoples from the mythical image of the Golden Age, Paradise Lost, and Christian millenarianism, to justify alien utopias. These exogenous representations first forge the image of supposed natural environmentalists without contamination from evil and then attribute to them the messianic task of "saving the planet" from the destructive action of capitalist modernity (Giraldo, 2014). These romantic idealizations are no more than fictions used portray externality with respect to global capitalism and build "the other" of modernity (Castro-Gómez, 2015).

Except for communities that remain in voluntary isolation, in the twenty-first century it is difficult to justify being "outside" of Western civilization. Local models coexist with hegemonic modernity that influences them and leaves tracks, or marks, of power on their bodies. That is why it is so difficult to talk about pure and pristine cultures. We can refer, however, to hybrid ontologies, composed of fragments, which are assembled among diverse traditions and heterogeneous worlds (García-Canclini, 1990). Tribes of modern people who intermingle with vernacular symbols. Cultural elements specific to the ontologies reproduced by global capitalism—a world made up of a collection of parts that can be manipulated through science, technology, and economics—and symbols where reality surfaced jointly in an inseparable link between the human and the nonhuman (Escobar, 2013). In short, we cannot refer to Western vs. non-Western dichotomies, as was the case until the great modernizing project of the twentieth century. Rather, we must speak of ontological hybridizations that manifest themselves fundamentally in rituals and localized practices such as agriculture.

Despite these observations, I stress that differences exist. Ontological differences are manifested in agricultural and aquaculture practices where rational discourse and productivist logic do not guide us regarding how to inhabit the world. These activities have been shaped by a history of collective knowledge and the deep bond that binds the everyday world of peasants, indigenous communities, nomadic pastoralists, and artisanal fishermen to the land. Although the ways of inhabiting the world of many of these people differ substantially from metaphysical economic rationality, this divergence is not explained by a simplistic dichotomy in which some are good and others are bad. It is a pragmatic rather than a moral question, learned through experimentation, trial and error, and an encounter with biophysical aspects specific to each territory, which has built up the knowledge base of many different people. Through nonformal—but necessarily empirical—learning, these peoples have understood that if natural cycles are transgressed, or if intervention is inappropriate at the wrong time, a decoupling of human activity and biological processes will ensue, preventing the material reproduction of family and community. Again, this is not the story of good green savages living in perpetual harmony, but the emergence of coevolutionary processes of reciprocal interaction between these societies and nature, whose learning permeated the way of understanding a world lived by the aegis of the organic and whose senses are coupled to the ecological conditions of the inhabited place.

For rural people who make agricultural landscapes, the way they perceive, feel, think, and situate themselves meaningfully in the world has been embodied in their daily activities. A very particular way of classifying and experiencing reality has been unconsciously internalized and imposed, although it is not always possible for them to translate it into a rational discourse (Leff, 2014). Rather, their reality is embodied in their acts, according to the historical conditions and the social relations of a human group that has created itself by transforming ecosystems and inventing habitats in the form of terraces, orchards, *acahuales*, *milpas*, or *camellones* (Giraldo, 2016a).

A good example of the continuity between being, doing, and knowing of the farmers of the Global South is their time horizon. For millennia, rural peoples have built a way of understanding and living life that continuously interweaves with cycles such as rain and drought, sowing and harvesting, winter and summer, the rising and falling moon, or the fertility of females for reproduction. This lived world is cyclical and nonlinear, because year after year, rural people know that nature will once again provide for the family. Their attachment to nature's cyclical basis creates a temporal coupling of human activity to these biological processes and allows, over time, material reproduction of their communities. Of course, given the hybridization of contemporary cultures, this cyclical time coexists in tension with the linear and infinite time of progress. Many of these communities have a constant relationship with modern markets, which creates syncretisms that are not easy to discern. Even so, in the gnoseology of many rural communities of the Global South, cyclicality continues to determine the productive activities and daily life (Ziga, 2013).

In these repetitive temporalities, there is no beginning and end between which "development" can be understood. There is a cyclical process that is not determined by any end point, with no undergirding teleologies or objectives. Circularity does not end in anything and does not evolve toward anything but its own continuation (Jullien, 2005). As in ecosystems, where all organisms produce waste that is food for others, so too the system as a whole does not produce waste, but is constantly feeding itself (Capra, 1998). Millennial Agri-Culture knows that forest humus enriches the soil and the manure of domestic animals is fertilizer for the orchard, in circular processes that neither begin nor end.

Polycultures are excellent agroecological tests of these ways of inhabiting domains and deriving knowledge through a pragmatic relationship with a lived world. This is knowledge co-constructed through collective experimentation, in which rural producers have learned to find virtuous associations and to unite mutually complementary parts, such as those found in Mesoamerican *milpas*, where corn is planted alongside beans, pumpkin, chili, and *quelites*. The knowledge that has emerged from the local base of natural bodies has avoided, through a daily relationship in constant search of "good living," the degradation of inhabited land, resisted disease, and provided a diverse and stable diet for the family. Furthermore, the organizational principles, the community paradigms, and the mutual exchange strategies that govern so many rural Latin American societies—such as barter and reciprocity in activities such as *mingas*, *tequio*s, or *mano vuelta*—were learned from the plurality, consensus, mutualism, and complementarity of polyculture. Agriculture, besides being a material source of production to ensure the reproduction of the family and the community, is the origin of cultural representations, cognitive apprehensions, identities, and collective meanings (Giraldo, 2014).

Peasant knowledge, regardless of ethnic origin, cannot be separated from its living contexts; it emerges in a continuous interweaving with the inhabited place. This interplay produces knowledge of the interrelationships between species occurring in productive diversity, cyclicality in the temporality immanent to the agricultural and temporal cycles, reciprocity in community relations, and the complementarity between the landscape and the areas of learning that arise from active participation with the inhabited territory. As phenomenologist Tim Ingold said (2000), this is wisdom that emerges according to how people find themselves situated in the world and is thus inseparable from the action and the natural contexts in which people participate.

A very important matter, which clarifies the point I'm trying to get to and that I take from Ingold's phenomenology, is that this knowledge that we reviewed consists of a whole body of knowledge developed over many decades and taught to every generation through pragmatic use. Peasant knowledge needs that daily experience, which is inseparable from the places where one dwells. This is knowledge that is completely dependent on its relationship with the environment. This is localized knowledge, which cannot be thought of in isolation from daily practice and experience of the world, insofar as—to paraphrase Maturana and Varela (2003)—there is a continuous coincidence between *being* a farmer, *doing* Agri-Culture, and *knowing* the lived world (Giraldo, 2013). What I stress about Ingold's argument is that peasant practices are not passed on from generation to generation, like the transmission of genes between parents and children. Rather, the relationships between farmers in specific environmental contexts support the continuity of a technical tradition. But not just technical traditions. Even the logic of peasant economies is not "inherited" independently of its application. This logic consists of a relatively autonomous flow of production and reproduction within the agricultural unit, in which one part is partially sold, while the rest returns to the cycle, in an iterative process that continuously returns to its point of origin (Ploeg, 2010).

It is at this level that I think that the fracture operates, because autonomous elements that pertain to technical traditions and peasant economies are intersected by heteronomous elements that change the context from which learning and daily practices emerge. If human action is seen as circumscribed in a permanent and mutually constructive relationship between people and the environment, as Ingold assures us, we need to see how peasants' experience and apprehension of the world change once the autonomy they maintained with their territory is eliminated and they are inserted into the global circuits of capital accumulation. In order to control nature, it is necessary for people to be colonized ontologically and epistemically: to erase the heterogeneity and diversity of worlds, to establish the only world of the Western homogenizing project. In other words, in order to subjugate and take over territories, it is necessary to modify the *being*, *doing*, and *knowing* of farmers, through the creation of populations that learn, think, and act, according to the economic rationality of Western modernity.

What interests me is how capitalism creates its own Other and actively fabricates it, converting chaotic diversity into orderly multiplicities, through the modernization of its practices and the generation of relationships of dependency. Further, I believe that this ontological and epistemic division of inhabited places achieves a much more effective and lasting process of dispossession than the physical expulsion of the inhabitants of usurped lands. Thus, the colonization of bodies is a requisite for the creation of long-lasting structures for land dispossession and the expansion of agro-extractivism on a global scale.

In Situ Deterritorialization of Peoples

We are trying to unravel the methods used by capital to separate the peasant from his land without physically displacing him or her. In this search, the notion of territory occupies a central place. Clearly, territory is not a materiality that can be understood independently of human intervention. On the contrary, it is intimately associated and in continuous activity with the various ways in which societies signify, perceive, and feel places. It is a hybrid, as Leff (2004) says, in which the symbolic is combined with the organic and the technological. This does not mean that nature and culture become the same thing. Rather, it is a game of relationships in which biophysical order is articulated with imaginary and symbolic order. A complex interweaving, in which the symbols of culture intertwine with nature; words intertwine with things; knowledge, techniques, senses, identities, and stories relate to landscapes.

In other words, we are undertaking a phenomenological reading of territory, in which we do not start from people on one side and places on the other, because as Heidegger (1994d, p. 137) observes, "Space is not in front of humans; it is neither an external object nor an internal experience. There are no humans and space," but both are in constant interrelationship. We do not imagine human beings detached from the world, but actively engaging in a process of habitation, where people, from the beginning, are in the world, affecting it and being affected by it. We speak of a territory from a non-positivist position, that is, by denying the assumption that places are there and subjects only go later to encounter them. On the contrary, territory is a network of dynamic interactions that arises in movement from our symbolic and biological tools that we create with others in society.

We are summoned to think about the place in interrelationship to bodies, languages, and biological and social history; always co-emerging, co-surfacing, as a result of the constant interaction between peoples and their specific natural environments (Varela, 2000). We deny the idea of ahistorical, prosocial, and pre-discursive places and instead prefer a concept of territory that comes from the embodiment of the Real, with collective histories, asymmetric power relations, and discursive processes of significance (Escobar, 1999). In summary, it is a question of understanding territory not as a thing, but as a relationship between materiality and immateriality, without separating place and population.

This phenomenological definition of territory helps us to build an expanded view of the phenomenon of land-grabbing, now re-conceptualized as territorial hoarding. As we have seen, this not only requires physical appropriation of space but also symbolic control. The territorial offensive and the deterritorialization of rural peoples that have intensified in the world since the dawn of this millennium necessarily require a combination of material and immaterial deterritorialization, of association between political-economic domination and cultural-symbolic domination, neither of which can occur separately (Haesbaert, 2011). I stress that in order to monopolize territories, it is not enough to have direct control over the land—the failure of the most scandalous attempts to dispossess land is proof of this. Much more subtle mechanisms are necessary, in order for power to discipline both space and farmers' bodies.

Deterritorialization has always been a process of physical and symbolic dispossession: of the inhabited biophysical space and perspectives of *being/doing/knowing* tied to these places. This was the case during the American conquest, the African colonization of the nineteenth century, the "peaceful" development project, and the territorial hoarding of our days. Capitalism creates its Other in two phases: first, by deterritorializing the existing forms of inhabitation, in order to later territorialize them according to its own rationality. The objective of separating the producer from the means of production—Deleuze and Guattari say—has as a requirement the destruction of territorialities that are later reincorporated as part of the system.

It is nothing less than a whirlwind that breaks up the interrelationships of space-time, changes places physically and symbolically, and mobilizes bodies to reincorporate them as functional pieces in the gears of production and value addition. Both for those expelled from their land and turned into agricultural proletarians and for those living in a place transformed by monocultures or who are inserted into a technological package for commercial agro-export, deterritorialization is an expression of the de-structuring of their worlds, modes of production, temporalities, meanings, and feelings, to be subserviently integrated into global agro-capitalism.

Inclusion and Removal of Autonomy

During a lecture on January 15, 1975, Michel Foucault (2014) mentioned two different types of exercise of power: that of leprosy—a system based on exclusion—and that of plague, based on inclusion. Lepers were banished and confined to areas isolated from all contact with the rest of society as a prophylactic measure to prevent the spread of disease to healthy people. The lazarettos—as these territories are called—were often enclosed by a barbed-wire fence to prevent the movement of sick people outside and inside their confines. Guards controlled the prison to ensure that the sick remained in confinement throughout their lives (Corzo, 2011). These were practices of marginalization, rejection, expulsion, and exile whereby an abnormal population was territorially encircled in order to purify society by severing relations with healthy individuals. In his lecture, Foucault argued that while the pattern of exclusion of the leper is a kind of power that continues to operate in institutions such as prisons or psychiatric hospitals, it is not the most widespread type of power in our contemporary societies. The model with the greatest historical fortune and which best explains the power relations of our time: cities suffering from a plague.

Unlike leprosy, in the plague-control model, there is no intention to exclude the sick, but rather to include them, with certain coordinates of control. The purpose is not to marginalize, but rather to incorporate the sick into a subtle epidemiological system that seeks to exercise power over their bodies. According to Foucault, when a French city was placed under quarantine in response to the plague, the city was divided into districts, neighborhoods, and streets. Each district and neighborhood had an inspector, and each street had a sentinel to guard each one of the doors. At the beginning of the quarantine, all citizens had to give their names, which were entered in a series of registers. Inspectors went from house to house every day to take roll call, and people were required to peer out a window when their name was called. If someone did not show up, it was taken for granted that he or she was sick in bed and therefore had to be treated. Foucault posited that such an organization is antithetical to that of leprosy. With the plague, the response is not to exclude or reject, but rather to establish a quarantine that includes people in a system, to establish not distance but closeness, and thus to determine whether individuals are adjusting to the rules.

I think the consequences of this power analysis are very different when the model is interpreted in the light of leprosy control or when analyzed from the perspective of the plague. A capitalist system that tends to generate exclusion and marginalization is different from a model that does not exclude, prohibit, marginalize, and repress, but on the contrary includes, incorporates, intervenes, and transforms individualities in accordance with its own project of political and epistemic organization. Criticism based on responses to leprosy leads us to conclude that a system that marginalizes should be replaced by an inclusive one. If the criticism comes from responses to the plague, a radically opposing corollary surfaces: the problem is not to include the excluded, but to create conditions that prevent their functional incorporation into the system.

Original accumulation can be viewed from the perspective of the plague as an incorporation process: the dispossessed were integrated as salaried labor into a system where, for the first time in history, the labor force was commodified. Now, that supposed excluded, expendable, excess population, often cited by certain left currents, is something quite different. It is a population that is included in the margins of the system. Those who migrate from the countryside to the city, who cross national borders, or individuals from the cities—not always from the popular sectors and some quite well qualified—who never manage to find employment, are, indeed, included in neoliberal economic globalization. In the margins of their apparent "exclusion," they play the role of keeping wages low, as well as being the excuse that allows employers to blackmail labor and be grateful for being exploited ("if you're not happy with your salary, there are lines of people who would work for half as much as you do"). The marginalized labor force also legitimizes the use of police by creating a sense of insecurity. Its role, in this latter aspect, is to ensure that "well-inserted" people do not ask for less, but more repression, more cameras at traffic lights, greater surveillance, and an increasingly closer watch by the powers that be.

It remains to be seen to what extent the newly dispossessed by territorial appropriation during this new millennium are being re-functionalized in this way. What is clear is that power is much more efficient when it includes rather than excludes. When power is imperceptible, when instead of feeling the oppression of direct violence, the so-called "excluded" believe that they are part of the forces of development. Here power is less expensive, more efficient, and long lasting. If we assume that capitalism is mostly an insatiable machine that engulfs the population to make it a part of its inner workings and that the function of development is to include by standardizing, to create uniformity, and to insert the populations subserviently to the US version of the Western project, then we need to focus on how processes of inclusion are defined by international institutions. It is the later that, ultimately, prefigure the mechanisms of power that are established in every country.

In the case of land-grabbing, organizations such as the FAO or the World Bank say that far from being undesirable, agricultural investment has been growing since the collapse of financialization, and thus, if they can, entities should take advantage of the opportunities that are opening up. The only requirement is to refrain from excluding the population and endeavor to include it. As an FAO report (2012a, p. 67) states:

> Large-scale corporate investment in agriculture can represent an opportunity. It can contribute to filling large investment gaps in poor countries with abundant natural resources but without the capacity to invest heavily in enhancing productivity. It can support the creation of infrastructure as well as the transfer of technology and know-how. Other potential benefits include the generation of employment and incomes as well as export earnings.

The discourse is based on the logic of incorporating peasant and indigenous communities through subcontracting systems such as contract farming, joint ventures, and business linkages between agribusiness chains and small producer cooperatives:

> Large-scale corporate investment in agriculture need not necessarily lead to the conversion of small-scale farming into large-scale agriculture… other more inclusive partnership models exist that are more likely to achieve desirable developmental objectives by successfully investing companies. In such models, local farmers would provide land, labor, and local knowledge, while corporate investors would provide capital, access to markets and technology and specialized knowledge. They would allow smallholders to make productivity-enhancing investment on their own farms. (FAO, 2012a, 2012b, p. 69)

The "inclusive agribusiness" models also consider leasing contracts—that allow large agribusinesses to lease land from small- and medium-sized farmers or to enter sharecropping contracts between agribusiness corporations and farmers—and linkages through cooperatives to connect to agribusiness and high value chains for export products such as biofuels:

> The misconceptions about biofuels are important to overcome for a farming community that has long suffered from low incomes. Bioenergy represents a good opportunity to boost rural economies and reduce poverty, provided this production complies with sustainability criteria. Sustainable biofuel production by family farmers is not a threat to food production. It is an opportunity to achieve profitability and to revive rural communities. (FAO, 2008, p. 97)

According to the neoliberal development discourse of these organizations, large plantations can offer advantages such as the construction of infrastructure, technology transfer, stimulation of innovation, increase in productivity, generation of jobs, and stimulation of growth. This is why the recommendation is that states should attract investors by offering "ease of doing business that favors the accumulation of capital in agriculture" (*Ibid.*, p. 59). This requires creating a favorable investment climate by ensuring that private property rights are respected, regulating institutions that enable markets to operate, encouraging dynamic input supply industries, expanding access to financial services for the implementation of productive projects, supporting agricultural research and extension, creating essential public services, promoting skilled labor, and facilitating access to technological packages including machinery, fertilizers, agrochemicals, and genetically modified seeds. The role of the state—says a most important agricultural development report written by the World Bank (2007)—is to correct market failures, regulate competition, establish public-private partnerships, promote competitiveness, and support the inclusion of small producers and rural workers in value chains for domestic markets and export.

The same report stresses that in this "new agriculture," rural communities must be offered opportunities to escape poverty through employment in agro-industrial enclaves and through their connection to high-value production lines. In this way, "peasant entrepreneurs" will become suppliers to modern markets and thus take advantage of the diversity of the rural world where small, medium, and large producers "live together harmoniously":

> Economic and social heterogeneity is a defining characteristic of rural areas. Large commercial farmers coexist with smallholders. This diversity permeates the smallholder population as well. Commercial smallholders deliver surpluses to food markets and share in the benefits of expanding markets for the new agriculture of high-value activities… The emerging new agriculture is led by private entrepreneurs in extensive value chains linking producers to consumers and including many entrepreneurial smallholders supported by their organizations. (World Bank, 2007, pp. 5–8)

Agricultural development based on large plantations with export-oriented technology packages requires education that stimulates entrepreneurial logic and the acquisition of technical skills in agro-extraction, so that "new entrepreneurs" can participate in the "opportunities" of emerging global markets. This is a strategy of incorporation, not exclusion, to be applied to almost half of the human population—three billion people according to estimates by the World Bank. An enormous project of agrarian geopolitics in which education is the political arm par excellence for mainly rural people to cease being what they are and join the world of neoliberal globalization.

Even so, the most important goal is not that construction of the Other, according to the principles of economic rationality, because in spite of its intentions, agribusiness education will never fulfill its objective of making people think in completely mercantile terms. The risk is not so much what pedagogy achieves in the promotion of its propaganda, but rather its ability to dampen creativity, collective imagination, and distributed intelligences. Heteronomy, understood as a regulation imposed by an external agent, is not as efficient at producing subjects as it is preventing alternatives from emerging. A calming and immobilizing system that is not as good at standardizing the human product as it is at numbing and brutalizing it.

The unspoken aim of the "new agriculture" is to deprive people of their own culture in order to westernize, standardize, and mold them into *homo economicus*. Since its inception, education, as understood by Esteva, Prakash, and Stuchul (2002) of Tolstoy, has been about "the conscious intention to transform someone into something" (*Ibid.*, p. 47). That something is the construction of passive, disciplined individuals who obey the wants and needs of capital. Entrepreneurial education of agroextractivism, akin to the colonial discourse of agricultural development of these organizations, consists of imposing discipline and control, as well as generating dependencies and incapacities. It seeks to transform farmers into obedient consumers of biotechnological packages and integrate them into specialized business agriculture, oriented toward maximizing profits and market dependence.

A developmentalist mindset, which the ministries of agriculture and rural development of the countries of the Global South follow with particular discipline, consists of modernizing the countryside, favoring large transnational capital investment, orienting production in accordance with comparative advantages and the vocation to export, and doing everything possible to subsume small farmers to a particular system of exploitation. We know the result already: greater dependency that translates into a sizable decrease in the options for small farmers (Ploeg, 2010). Once the life of rural communities has been integrated into the global movement of goods, they become vulnerable to drops in prices, increased costs of trade inputs, tax increases, phytosanitary and animal health risks arising from the introduction of monocultures and commercial animal species, and the leonine clauses of association contracts.

Since its inception, rural development has been an excellent means of intervention in the life of rural peoples and, with the promise of aid and higher expectations, has succeeded in creating incapacitating dependencies and a system of needs protected by modern economic rationality (Illich, 1996). By means of dispossession, "ontology-cides," and "episteme-cides," rural development has been transforming people into clients of services—such as rural extension and technology transfer—once offered by governments, but now delegated to private companies who "accompany" "business farmers" allied with large agroextractive companies. This is where the new agricultural geopolitics is changing. The activities that government programs had undertaken since the 1970s to bring the green revolution to peasant families are now being transferred to private investors, who become "companions" and "partners" of the "new entrepreneurs." This is how the costs of expanding industrial agribusiness are reduced, thus helping to widen the reproduction of capital.

By entrusting the task of reproducing capital to individual capital, large money interests encourage private agents to make decisions regarding what, how, where, and when to plant and market. In the end, this immense power translates into a form of territorial control, because the control of large areas of land accrues to big agro-industrial entrepreneurs. That is why I insist: Land dispossession must not be defined as the simple monopolization of land by private investors, but rather should be rethought as a holistic form of territorial control by a handful of private capitalists who collude with national states and multilateral organizations. Because, as Marx suggested in his letter to Vera Zasulich, "in order to expropriate the farmers, they must not be driven off their land." In reality, to monopolize territories, it is not enough to have direct control over the land. Sometimes it is much more effective not to dispense with local communities, but to integrate them into the mindset of agro-industrial plantations, which requires deterritorializing their knowledge and reproducing a discourse coupled with the dual ontologies of modernity.

With "contract farming," capital also solves the problem of the workforce. Behind the promotion of "partnerships" between agro-corporatism and small farmers, a hidden agenda stipulates that the former will gain greater value from the latter. Thus, farmers who previously were independent become dependent on the agro-industrial chain. Because agreements between the parties are asymmetrical, companies promise to provide technical assistance and manage loans and subsidies, so long as producers sell their output to them. Small farmers assume all the risk, become indebted, and endure discounts in their pay for input costs and technical assistance. Companies, on the other hand, are undisputed winners: they indirectly profit from the work of others without granting any compensation and ensure a supply of raw materials they need for the production of their processed foods without ever working the soil (Dusch, 2017).

The biggest experiment in this strategy of territorial appropriation through inclusion is the initiative called the World Economic Forum's New Vision for Agriculture. This is a "market-based" program that, until 2017, operated in 21 countries in Africa, Asia, and Latin America and mobilized $10.5 billion to involve 10 million farmers. The program, in which the G7 and the G20 participate, is led by 31 of the largest transnationals in the world[7] and aims to develop new models of partnerships between corporations and farmers.

The exercise of power over territories without having first expelled its residents is what Rogerio Haesbaert (2011) calls deterritorialization in immobilization. It arises when inhabitants do not control their territories, but rather are controlled by others. Thus, a deterritorialization without physical displacement occurs, since heteronomous elements fracture the decisions and specific ways of inhabiting rural communities. When people are incorporated into the global flows of capital accumulation, they, the object of this strategy, lose "control over their territorial bases of reproduction and reference" (2011, p. 211), thus losing autonomy of their own lives by ceding "control over their reproduction space." The new geopolitics of agricultural development seeks to include all possible territories into worldwide capital accumulation,

[7] A.P. Møller-Maersk, A.T. Kearney, BASF, Bayer, Bunge, Cargill, Carlsberg Group, CF Industries Holdings, Deloitte, DuPont, Heineken Global Supply Chain, International Finance Corporation, Louis Dreyfus Company Asia, McKinsey & Company, Mondelez International, Monsanto Company, Nestlé, Novozymes, PepsiCo, Rabobank International, Royal DSM, Sinar Mas, Agribusiness & Food, Swiss Re, Syngenta International, The Coca-Cola Company, The Rockefeller Foundation, Unilever, UPL Limited, Wal-Mart, Wilmar Investment Holdings, Yara International (World Economic Forum, 2017).

which means, in the words of Haesbaert, a "deterritorializing territorialization," since in the end the peasant is dissociated from his land, after big money interests master his practices, knowledge, fields of enunciation, and position in the world.

The necessary condition for the expansion of capital is the separation of peoples from their material and symbolic conditions of existence (Porto-Gonçalves & Leff, 2015). However, this separation is triggered both by the physical displacement described earlier and by the imposition of a regime of truth, which reproduces certain knowledge and certainties, while excluding all discourses and practices that are counterproductive to the regime of the truth of agricultural development.

Control of Enunciation Fields

> Seventy-two thousand four hundred repetitions create a truth. Idiots!
> Aldus Huxley

The interpretation of territorial control cannot dispense with the role of language in the configuration of "worlds." In fact, we begin by abandoning the notion of pre-discursive places in order to approach the analysis of territoriality from a perspective that includes the fields of enunciation. We need to highlight the effects of language on the perception of reality. As the phenomenological, hermeneutic, and post-structuralist schools have noted, the fact of apprehending and perceiving in a certain way, and not another, is linked to the possibilities of the language to which it belongs. We humans have this resource to express one of many possible realities. As Wittgenstein noted (1988), we are predisposed to think, perceive, and even to feel in a particular way, according to inherited imaginaries by belonging to a specific language community.

Language is a mediator between the world in which we live and share with others in society and ourselves. We do not become aware of the characteristics of the environment and communicate them objectively, because language does not denote an objective reflection of reality, but rather is a tool we use to creatively project "worlds." It does not copy that which is independent of the subjects, but rather is a mediator that "builds realities." Language has this enormous capacity to mediate between our body and the environment to observe it in a particular way and to make realities "arise" that we bring as firsthand experiences of the world (Maturana & Varela, 2003). For this reason, we ought to pay attention to that creative dimension of language, in which "by saying, things get done"—as Austin noted—as in the projection of images that are evoked by words to appear in collectives.

As we saw in the previous chapter, the discourse of development is permeated from beginning to end by an anthropocentric economic rationality, based on rational subjects that manipulate inert objects according to the needs of production and consumption. This discourse represents all sectors of the world in terms of productivity, performance, profitability, efficiency, or utility, re-signifying symbolic orders and life itself, according to the criteria of economic codes (Leff, 2004). Rural benefactors of "institutional support" are renowned as "entrepreneurs," "small and medium producers," or "agricultural entrepreneurs" in development agency projects and are carefully diagnosed and analyzed in indicators of "competitiveness," "productivity," and "profit." Also nature, in this particular way of constructing reality, is called a "natural resource," "natural capital," "genetic resources," "plant material," "environmental services," "natural goods or assets," or "commodities" and is referred to as the symbolic-biotic relationship with expressions such as "exploitation," "use and control," "management," or "administration and management." These are all examples of how in the language of development nature takes the form of an object and a commodity and how our relationships with each other are expressed as transactions.

This is no small matter, because words are not neutral. They bear marks of power that lead the partners to project "one world." This type of discourse opens the imagination and drives perception in accordance with modern images that separate culture from nature and express the dominance of human beings on earth. Their semantic charge treats ecosystems as simple objects that can be made available, economized, and capitalized (Noguera, 2004). In this creative capacity of language, the figure of the metaphor occupies a fundamental function. This is because the metaphor, instead of saying, "this is like that" by way of analogical comparison, indicates "something *is* that" (Ricoeur, 1980). This occurs when we hear that nature *is* a "resource" or that the peasant *is* "a small business farmer." These metaphorical statements are not simply ways of speaking, but powerful policy instruments that create images, "make people see," by establishing economistic meanings and commodifiers of the world.

When the discourse suggests, for example, that we do not breathe, but receive an environmental service, a modern regime of truth about nature is actually being built. Because little by little, by repeating it, the economistic metaphors become a common way of speaking, shaping a metaphysical way of affection and perception. When they intervene in the lives of people, these metaphorical statements generate perceptive beliefs and reproduce power by forcing life and work to be seen through the lens of economic rationality. The dogma of development transgresses the life of agrarian, pastoral, and fishing communities, evoking images from modernity through metaphorical expressions, which, when repeated constantly, become indisputable axioms.

The technological, cultural, and representational regimes associated with the discourses of "new agriculture" have created semantic scars between those who accept and reproduce them, since the metaphors that surface from economic rationality, when adopted and circulated by the population, reproduce the linguistic regime and the system of truths needed by agroextractivism to maintain its hegemony. More empirical information is needed to understand if the communities that have assumed the above metaphors in their narratives feel that water, land, minerals, or forests are a quantifiable asset for meeting the needs of humanity or believe that the human species has the world at its disposal. My provisional answer is that people are not passive agents and do not immediately lose sight of their existence solely by repeating metaphors of modern ontologies or joining the market. Yet, we ought not to fool ourselves. It is very evident that rural people are increasingly becoming more modern.

Even so, we still have to go further. In reality, economistic statements cannot be imposed as if a whole body of meaning could be ideologically transmitted to people independently of the contexts in which they live. Rather, these verbal conventions are incarnated and become permanent dispositions (Bourdieu), once communities find these metaphorical statements congruent with their own experience. To embody a discourse, as Ingold (2000) points out, it is necessary to make comparisons between their own daily practice, their sensory experiences, and all the linguistic constructions that end up affecting the perceptions of peasants about the world around them. The perception, says Ingold, is the result of an encounter with the world and from the world, a process of active participation with the environment, where what is perceived is perceived according to contextual conditions and specific environmental characteristics. As phenomenologists insist, apprehending is not about taking a view "about" the world, as if it were "out there" and we had to parachute into a reality that precedes us (Varela, Thompson, & Rosch, 1997). Apprehending, on the contrary, means getting involved, dwelling, forming a vision "of what is inside" (Ingold, 2000). We live in a world with others, so our way of *being* and *inhabiting* surface according to a history of relationships between the human body, language, and a lived environment. Our perception appears according to the place and the social relations in which we experience it.

To know the world implies discovering it in a direct way while the body moves in a specific environment (Ingold, 2000). Thus, the discourse that is trying to be imposed on rural inhabitants is not acquired in a passive way: It is tied together and emerging continuously according to the spheres of life in which these speakers participate. The nature/culture separation that underlies agronomic and zootechnic practices, coupled with the dichotomous discursiveness of rural development, can only be assimilated when living in a world dominated by these dichotomous meanings in a practical and permanent way, such as when one perceives them from uniform plantations of palm, soya, sugar cane, or grassland. It is impossible to think of *knowing*, *doing*, and *being*, independently of the contexts in which people find themselves. What people perceive of their world, and what they call it, will depend on how they engage with the environment surrounding them.

What I would like to emphasize is that modern stories are only embodied when they are lived daily in a way that is coherent with the meanings of these stories. In other words, the metaphorical expressions mentioned above can be embodied when, for example, farmers are no longer able to reserve the best part of the harvest for the next planting season. Or when they assume tasks associated with the extraction of nutrients, such as the use of highly mechanized technology, directly sowing transgenic seed and dousing them with chemically synthesized fertilizers and pesticides. The division between nature and culture that underlies the economic stories of rural development is internalized in the bodies of the population that is the object of modernization, once people assume the practices and knowledge typical of agroextractive modernity and live in a world dominated by the geometric esthetics of monoculture. So the metaphorical function of the language in which things are done by talking, such as making nature "look like" an object to be exploited, and making life itself look as if we were all traders carrying out transactions in the marketplace, motivated solely by the desire for profit, becomes stronger when the deterritorialization of life worlds is tied to this particular way of enunciating the world.

In the end, as Foucault suggests, power is much more effective when it is not coercive, but when it relies on the creation of common sense. This also occurs when it creates certainties, rituals of truth that are accepted as unquestionable, and when the world of everyday life is subjected to certain orders of discourse. In the case of agroextractivism, the globalization of modernity is based on an economistic logic that permeates the entire language. As Porto-Gonçalves (2006) says, this logic both divorces peasants from nature and separates them from each other, individualizing them, fracturing their community networks, depriving them of the autonomy they maintained with their territories, and introducing them into the ontology of commercial competition.

Agribusiness and Territorial Rent

Until now, we have followed the thesis of incorporation from the active creation of the Other and the production of subjects related to economic rationality and the order of meanings that structure the ontology of modernity. It is the classic developmental strategy to universalize the modern project on a global scale and to beget the Other from sameness in ways that are useful for reproducing capital. Let us recall the argument: Capitalist growth does not have the luxury of exempting any territory from its dynamics of valorization, and that is why it is important to include everything lying in its path within the system. In agriculture, however, this objective is more difficult than it seems, as it is practically impossible to monopolize the entire world's land to transform it into uniform agro-industrial plantations. Insofar as there is land with low fertility, difficult access, and precarious infrastructure, it is very costly for financiers to invest in intensive monocultures directly. It is more feasible to make available to capital the land on which millions of people cultivate indirectly.

For this purpose, the farmers of the Global South are indispensable, not only because they are the ones who cultivate and whose animals graze on land with these characteristics, but above all because by joining together they are an economic force of unfathomable dimensions. A peasant in Nyéléni, Mali, said it accurately: "We peasants are the world's largest investors" (LVC, 2015a). The data back him: Together they own more than half of the world's arable land, that is, an area estimated at 764 million hectares, more than the land owned by large landowner agrocapitalism (ETC group, 2009). In other words, together they have a greater economic capacity than any transnational firm does and hence the need not to exclude them, but to incorporate them into globalized capitalism. Yet if it takes so much work to standardize them through the developmental enterprise, there has to be a less descriptive and more theoretical explanation for this tedious process.

A good answer lies in the analysis of rent in the third volume of *Capital* and revisited by Armando Bartra (2006). According to Bartra, the survival of the peasantry, in spite of its predicted extinction, is because capitalism could not exist without peasants, a conclusion he reached when analyzing territorial rent. The system depends on them, not only because they provide cheap, seasonal labor for commercial agriculture,[8] but above all because in agriculture—like all other activities that depend directly on nature, such as mining or oil extraction—there is a matter called differential rent.

Bartra's explanation can be summarized as follows: Agricultural activity is based on land, i.e., a scarce natural asset with different qualities of fertility and location. This characteristic means individual capitalists monopolize the best land, while less commercial land, more barren and difficult to access, is left to small farmers. Thus, an agrarian structure exists in which large landowners coexist with medium and small farmers, all of whom contribute to the market with different productivities. However, to the extent that even the worst soils are required to make a profit or at least recover costs, the price of agricultural products is not defined around average production costs—as is the case in industry—but based on the costs of less fertile soils that are in a worse location in terms of the market. This way of determining the characteristic price in agriculture means that society has to pay a surcharge, an extra payment, for agricultural products, which is shared among the capitalists who monopolize the best land once they have recovered the investment and obtained an average profit (Bartra, 2006).

What Bartra wants to ensure is that this "extraordinary profit," or rent, which is appropriated by the agro-capitalists, comes from the common fund of global capital, which, in turn, cannot originate from any other source other than the surplus produced by labor. We all pay rent as workers and consumers, by forfeiting a premium for food, a kind of tribute that is distributed among the landowners who own the best land. This is mainly the reason why capitalism builds a dual agrarian structure composed of capitalist units—large and medium—and peasant units, i.e., there is a need for the latter to contribute their production to favor the appropriation of the differential rent of the former (Bartra, 2016). However, there is a detail. The agricultural investor is not served by an autonomous peasant who, working from his own rationality, supplies food from his surplus crops to be sold in local markets. On the contrary, it needs a modern and commercial "entrepreneurial peasant farmer" who produces the same products but with much lower yields so that the investors' extraordinary rent can last and increase.

This is a good way of understanding why states prioritize "production lines" or "product systems," oriented toward agro-export, and explains why agricultural policies are making every effort to create linkages where small, medium, and large

[8] The peasant sells only part of his labor force because it is only a supplement to his income as a direct producer, and therefore he is willing to work for a lower wage. Without peasants, no one would be willing to work solely during the harvest season, and society as a whole would have to pay the remaining income needed for the subsistence of the seasonal employee, so pure capitalism is impossible in agriculture (Boltvinik, 2009).

producers live together "harmoniously." The objective is to set in motion a maneuver of indirect dispossession, exploiting the work of others through their submissive incorporation into neoliberal globalization to appropriate the differential rent. Furthermore, speaking of rent, we are widening our understanding of the concept of "accumulation by dispossession" coined by Harvey, because dispossession, as Bartra notes, is not accumulation in itself, but only the premise required for accumulation.

In agriculture, as we have seen, dispossession is taking place through the theft of work; through mechanisms such as contracts of association, the conversion of peasants into small agro-entrepreneurs, and the creation of new clients for the inputs and services of the agroextractive model; but also through the commercialization of seeds, knowledge, and agro-biodiversity. Thus, resources are made available to capital that cannot be produced by capital on its own account. However, as Bartra (2016) says, what begins with dispossession culminates in the valuation of what is dispossessed, which does not occur in terms of productive investment, but rather through the monopolistic ownership of scarce goods whose exclusive ownership is subject to speculation. We see how the valuation of the expropriated occurs through the appropriation of differential income and the creation of consumers of products from the petrochemical industry and financial services. But also, as we saw in the previous chapter, the crisis caused by the model is a good opportunity to increase agricultural variety used in the agri-food, cosmetic, and pharmacological industries; the expansion of the markets for organic industrial products; and the environmental markets for the so-called "ecosystem services" of the new geopolitics of sustainable development. With all of these different operations, we speculate on common goods, privatized without retribution to those who were dispossessed.

In terms of agro-biodiversity alone, farmers, pastoralists, and fisherfolk are crucial, as these populations raise 40 species of livestock, plant some 5000 species of crops, and fish 15,000 species of freshwater fish (ETC group, 2009). This enormous diversity is necessarily tied to the knowledge and practices of people in their age-old processes of ecosystem transformation. Capitalism has become aware of an untapped wealth, and thus its undeniable interest in integrating these common goods into bio-trade by means of their articulation with business economies. Capital does not always transform the Other into its own image and likeness. Sometimes it is more useful to employ a pre-existing exterior, incorporating it and re-functionalizing its difference. In the case of indigenous peoples, for example, the discourse of the good green savage has evoked an idyllic, bucolic image, which has been used for ecotourism investments and emerging green markets (Ulloa, 2004). Agroecology has also been the subject of co-optation by the institutions that govern agriculture around the world, which highlights the growing interest in not wasting the traditional knowledge built over millennia by farming communities.

Territorial control through the mechanisms described herein is an excellent business for a capitalism in crisis that requires land in order to obtain the returns it is not finding in other productive sectors or in financial speculation (Bartra, 2013). The extraction of minerals, oil, wood, food nutrients, agro-energy, and, soon, water is a respite for a sick capitalism that urgently seeks to control all possible territories, for which—as I hope to have shown in this section—it does not always need to displace local communities. On many occasions, it is more useful to integrate all sectors of the population into a renewed interest in incomes. This inclusion of rural territories previously marginalized from the dynamic of capital accumulation will increase the diversity of returns and, thus, differential rents. As Bartra assures us, income depends on the scarcity of what we want to monopolize, so the predictions of scarcity increasing into the future are excellent news for the plans of speculative capitalism as it seeks to avoid the downward trend of profits and volatility in the stock markets.

Nonetheless, some of the pieces of the puzzle are still missing. We need to analyze further the way territories are dominated by controlling bodies. Already we have seen control over the fields of enunciation. Yet a more exhaustive treatment of the ontological experience of power is still needed, since it is inscribed in the bodies and the government of affections, not only of farmers but also of the population as a whole that depends on the agri-food system.

Chapter 4
The Government of Affections

> By affect I understand affections of the body by which the body's power of acting is increased or diminished, aided or restrained
>
> Baruch Spinoza

Our discussion until now is based on one of Marx's formulations, according to which the secret of primitive accumulation lies in the separation of producers from their means of production. We take seriously, however, his clarification that carrying out this uncoupling does not always involve expelling inhabitants from their lands. Based on these two premises, we posit that, more often than not, separating people from their territories of life occurs not through exclusion but rather by expropriating people's conditions of autonomous habitability and incorporating them functionally into the system. This happens through multiple mechanisms of deterritorialization grouped under the name of "development." In the case of territorial hoarding, for capital it is both more efficient and less risky to control certain territories indirectly, through the usufruct of labor, wisdom, and ecosystems, by promoting neoliberal agriculture and integrating territories into the geopolitics of sustainable development.

In this chapter, we will continue to reflect on these ideas based on the arguments presented hitherto, but focusing on an aspect that deserves particular care: capital accumulation through affective separation of bodies from their living spaces and from all other human corporalities. We posit that in order to appropriate agricultural rents, accumulation by dispossession, in addition to what we have discussed so far, requires reorienting rural inhabitants' affective relations and sensitivities among themselves and in relation to their spaces of reproduction.

In other words, valorization of capital rests on, first, a rupture that decouples a certain affective relationship that ties rural producers to one another and also ties them to the land; this is followed by a reorganization of their behavior, emotions, desires, and feelings in a metaphysical imaginary of production and consumption. I propose to focus on the affections as the space where power circulates and is

O. F. Giraldo, *Political Ecology of Agriculture*,
https://doi.org/10.1007/978-3-030-11824-2_4

exercised. In particular, I am concerned with how power relationships are interwoven between agroextractivism and the perceptible order that guides human experience.

I begin this section by continuing the phenomenological description that I outlined in the previous chapter regarding how agricultural producers in the Global South relate to their habitation units and then discussing the dispositive of political subjugation based on capturing their affections. Based on the phenomenal assumption that our encounter with the world is always situated and embodied, I argue that territorial control cannot exist without being inscribed in the body and the feeling horizons of those who are subjugated. I wish to explore the hypothesis that the efficacy of such an affective conquest resides in the esthetic. In our case, this is expressed as a technological ability to produce ordered space according to the orthogonalization of the agroextractive project. To conclude, I will insist that there is no deterritorialization that is not influenced by the affective flows and the emotional regulation of the subjects and by the production of spaces in accordance with the geometrization and disciplination of nature.

Creation of Linkages with the Place and the Political Administration of Affections

Much is said about the transcendental change that occurred when hunting-gathering-scavenging nomadic societies became sedentary societies in the Neolithic period. This is usually mentioned as a way to trace the history of anthropic ecocide that has led to our current environmental crisis. I would like to delve into this ontological transformation to undertake a very brief affective archeology regarding the place and then explore the split between rural peoples and their territorial foundations of social reproduction.

My thesis is the following[1]: The invention of agriculture made it possible for constantly moving nomadic groups to opt for a life rooted in a space that is not abandoned, for a dwelling that no longer is constantly moved. This fundamental transformation meant that those human collectives began to feel as though they were "people of a place." To become sedentary meant that the human animal found a permanent place to live and, thus, life from then on out became specific to the place chosen to live. With the creation of agriculture, human habitation is located next to crops and domestic animals, so that, slowly, as Saint Exupéry's "The Little Prince" says, places are domesticated in order to "create linkages." Those spaces create a sense of belonging that is different to that of nomadic peoples. For this reason, since then and until very recently, sedentary humans being thinking of the place they inhabit as different from any other. It became the homeland and home. It is the only place where people feel they belong.

[1]The theoretic background of this first part of this chapter was thoroughly covered in my article Giraldo, O.F. "Hacia una ontología de la Agri-Cultura en perspectiva del pensamiento ambiental", *Polis Revista Latinoamericana*, 12(34): 95–115. 2013.

With agriculture, these spaces are qualitatively different from the surrounding land. Through the permanent habitation of societies linked to specific places, those spaces become depositories of histories and symbols while becoming sacred, i.e., they become territories. With the agricultural revolution, ecosystems are radically transformed (often linked with the beginning of the sixth massive extinction of biodiversity), but also, with the creation of agriculture, human beings become ontologically distinct from the preceding nomadic hunters. As Mircea Eliade (1981) says, pre-agricultural communities could not feel the same way or with the same intensity, grounding, and belonging to the land. Thus, a culture in transformation had to replace the symbols of a hunter-gatherer society with those having to do with sexuality, fecundity, the sacredness of the woman, and the earth. The symbol of Mother Earth that emerges with agriculture clearly expresses humans' feelings when they acknowledge their relationship to the earth. The religious experience also becomes much more concrete: It blends intimately with the seed, the earth, and the rain. It merges more completely with life.

Much more than the economic sector touted by modern economic rationality, Agri-Culture is a foundational expression of an ontological transformation. As humans we conceive of ourselves tied to concrete places, understand ourselves as belonging to a place-territory that produces affections, feelings, sensations. In other words, our acknowledgment that we the children of sedentary inhabitants of the land is likewise an acknowledgement of an originating affective condition of beings affected by belonging to the earth. This involves recognizing the intimate link that unites the *being* of someone who is an Agri-Culturalist with the place in which crops have been sown. It is a relationship with the land mediated by work, involving wisdom, practices, feelings, sensibilities, and affections. "Create linkages," as The Little Prince taught, is an empathetic affection. It is an affective and affecting relationship, in which both the agricultural producer transforms the ecosystem and the producer is transformed by the land on which he/she toils. This is a bidirectional relationship whereby the Agri-Culturalist inhabits the plot of land while, at the same time, the plot of land inhabits him/her through a repertoire of symbols, rituals, and affections.

Yet the affective relationship of belonging to a specific territory also links those who acknowledge themselves as being children of the same place. Communities of fellow citizens have known how to live among themselves by building bridges, though never free of tensions and conflicts. They constitute a social life where, through a network of relationship, life can be lived well using strategies such as "don", kinship, reciprocity, festivities, and collective work. These are inter-corporalities that do not depend on egotistical volitions or to the "I" of modernity[2] (Varela, 1998), completely outside the de-territorialized life of our contemporary world. The collectives that

[2] Charles Taylor says in this regard: "This is something completely new in our history, of being able to say in the past two centuries "I am I." Previously, we did not use the personal pronoun *I* with the definite or indefinite article—*the* or *a*. The ancient Greeks or Romans and the people of the Middle Ages never used them as a descriptive expression. Now it is possible for us to say, "There are 30 people, or *I's*, in the room," but our ancestors would not have said it the same way. They would have said, "There are 30 souls in the room," they would have used any other descriptive term, but they would not have used the word *I*" (Varela, 1998, p. 21).

descended from these communal forms of habitation affected the ecosystemic order with their common knowledge, while also being affected by the creation of inhabitable habitats. Bodies-plots, bodies-knowledge, social-bodies. In other words, relational ontologies in which the "being" of many people of the global South cannot be conceived as independent of their territory and the immanent communal structure of how they practice their lives (Escobar, 2015). These are forms of existence without ruptures between culture and nature, or between the individual and the community.

What I would like to stress is that this historic overview is the affective base that unites the body to the place and a set of bodies to one another. Without this myelin that constitutes empathetic affection, there is no grounding to the place nor sociality. Bodies are detached from the land. Bodies separate one from another and become independent "I's." It is precisely here where dissociation occurs and where the power of development passes through, i.e., the empathetic break that estranges people from the place and breaks community ties (Toro, unpublished). Linkages are broken, and belonging to the land is forgotten. Marx was right. Capital accumulation is based on a separation, i.e., the one involving human bodies and their livelihoods. What he failed to mention is that this division requires intervention in the order of affections and feeling patterns regarding their living spaces. This implies the establishment of a perceptible regime that excludes attachment to the place and to cooperative communality, and which dis-incorporates people from the abstract symbols of economic value.

I revisit the notion of "economic rationality" that I have used interchangeably as a heuristic concept to bring greater clarity to the process of fissure, which I take from Marx and hope to explain. I have used this expression to name a particular way of signifying the world according to economic science and the symbolic suppositions of Western modernity. This "rationality" is not used as a synonym of reason, but rather as a term associated with the way we humans affect and are affected by reality. With this definition, I hope to avoid the Cartesian dualism that divides the world into two categories, the mental and the sensitive; the *cogito* as immateriality as opposed to the universe of no-reason (Rorry, 2009). We began, however, from the fact that we cannot exist without a body and thus the absurdity of fragmentations between heart and mind, disjunctions between emotions, feelings, and affections, and what we abstractly call "reason" (León, 2011). Perhaps we are rational beings—it seems that debates have not closed—but this assertion makes no room for divisions between the mind and the emotional world and even less space for a disembodied analysis of human ontology. Whatever way we address rationality or irrationality, including "economic rationality," it should be done based on a feeling and affective body that thinks and inhabits a flesh-and-blood body, without anatomical dissections or Platonic divisions between the perceptible world and the world of ideas.

To address "rationality" as inseparable from the affective order also implies accepting that the act of "*sentipensar*," [thinking-feeling], is the outcome of interactions between phylogenetic history and cultural history, or what Varela (2000) calls the result of interdependencies between biological baggage and social conditioning. Logically, like other living creatures, we are predisposed to feel according to the specific bodily characteristics of our species, but in addition, as human animals we "think-feel" according to the concrete feeling constructs of the societies to which we belong. As Emma León (2011) wrote, although emotions such as joy, sorrow, sadness, fear, or anguish seem to be individual, original, and innate ones, they are in fact influenced, conditioned, and unfold in keeping with the environment of social signification in which we find ourselves. Not only do we feel—not even rationalizing—according to the characteristics of our personality, we are inclined to do so according to the cultural shaping of the communities of which we are members.

By accepting that an emotional structure is an eminently social affair, we accept that that it is intersected by power relationships. The notion of *ordo amoris* that León borrows from Max Scheler says a lot in this regard. Scheler understands *ordo amoris* to be "the order of things that can be loved and, thus, things that cannot be loved" (2011, p. 40). This corresponds to a distribution of feeling that organizes the experience of subjects, determining possibilities of being affected by some things and not by others. *Ordo amoris* watches over those aspects of the world that can be felt and establishes before other aspects reacting in an insensitive way. Thus, we can deduce that economic rationality, or the economistic *thinking-feeling* that seeks to impose a large-scale development project on the people of the Global South, is the *ordo amoris* of homo economicus. We are not referring to massive anesthesia or a de-humanization. It is rather more a reorientation, a channeling of the stimuli of the physical world in which the subjects can be feeling beings. The meaningful and evaluative elements that are of use to the dominant force are incorporated in dominated subjects in this sensitive regime.

We human beings cannot live in permanent analgesia. But the order of affections creates a frame of reference regarding that which we can really feel. Under the perceptible regime of economic rationality, the sorrow of the mutilation of the land cannot be felt as sorrow. Since the heart and the passions are programmed to *think-feel* of nature as a stock of inert resources, the affection for the homeland, inherited from the agriculturalists who had converted themselves into "people of the place," is transformed into love of a heroic feat of conquest. The footprints of war against nature, the ruins, devastation, and desert created by the Green Revolution are not a part of the reign of things that can be felt as sorrow. Quite the opposite. The perceptible regime creates a drive to occupy spaces, an affection for an agro-biotechnological victory over living forces. This is dis-empathization of the scars left by an agro-industrialized geography erected in the name of economic progress. A fundamental fracture in the "linkages to the place" and the "place as a homeland" has occurred, and de-territorialized linkages have been created, emotional links to other aspects provided by the market. In the final analysis, as Foucault suggested, the effectiveness of power rests not in coercion, but rather in the logic of common sense regarding the content and values that subjects adopt from the world.

In *Brave New World*, Aldous Huxley saw this very clearly by showing the effectiveness of a government that operates in the interior of human beings. This is the vigor of a power that lodges in the bodies and organizes the affections and desires of the population. Once it is able to govern the aspirations, dreams, and certainties of the governed and, in the final instance, their happiness, then no physical violence against people is needed, because the objective will be fulfilled insofar as they will love their status as servants. Reinterpreting Gramsci from a Foucault perspective, hegemony is not achieved by obligation or coercion; it is much better to do so through the skill by which a social class imposes its "worldview" on the rest of the population through seduction and persuasion. Hegemony, seen in light of Scheler's *ordo amoris*, is the ability of the dominant to bring about an ordering of the body and affections by building a "regime of affectations," with which to accumulate capital through the accumulation of bodies (Castro-Gómez, 2015).

In the case of territorial control for the appropriation of rents, a direction for affective structures is also required, i.e., a molding of the emotional repertoire and a production of subjectivities. This involves hording of territories-body, plots of land-bodies, knowledge-bodies, and social-bodies using affective capture mechanisms. This requires political administration of the corporalities lending direction, meaning, and content of hegemonic economic rationality to the experience. In addition to a utilitarian and objectified vision of places, agro-capitalism needs a vision of *itself*. A self-perception of self by rural inhabitants, whereby, in addition to seeing themselves decoupled from the rest of their colleagues, and estranged from the inhabited land, they self-conceive of *themselves* as resource intermediaries, as "entrepreneurs" of themselves, and as administrators of their own "human capital" (Ibid., p. 378). If, through the creation of agriculture, a powerful ontological transformation had occurred, with the territorial growth of capitalism, another ontological transformation happened, i.e., the affective de-linkage of the place and the reorganization toward an emotional attachment that is de-linked from the original belonging to the land and human sociability.

Foucault (2002) gave this process the name biopolitics: a technology tasked with regulating life, of actively "fabricating" populations such that their entire lives are at the service of capital. Agribusiness uses biopolitics as a way of exercising power in order to governmentalize intimacy, administer corporalities, and produce docile subjects separated from one another and unleashed from the land. Its objective in this case is to grow geographically by dominating bodies, appropriating territories through a government of affections. The conquest of bodies is the fundamental tool to activate an agro-industrial locomotive, i.e., a model of death and desolation that channels people's desires, sensibilities, and affection in the *ordo amoris* of productivity, profitability, and the market by transforming them into functional instruments for the appropriation of territorial rents.

Constructing a Shortage and Producing Desire

> The discourse —psychoanalysis has shown us— is not just simply that it exposes —or hides— desire; it is also the object of desire.
>
> Michael Foucault

An African woman bent under the sun, weeding sorghum in an arid field with a hoe, a child strapped on her back—a vivid image of rural poverty. For her large family and millions like

> her, the meager bounty of subsistence farming is the only chance to survive. But others, women and men, have pursued different options to escape poverty. Some smallholders join producer organizations and contract with exporters and supermarkets to sell the vegetables they produce under irrigation. Some work as laborers for larger farmers who meet the scale economies required to supply modern food markets. Still others, move into the rural non-farm economy, starting small enterprises selling processed foods. (World Bank, 2007, p. 1)

This image proffered by the World Bank is compelling: This peasant woman is suffering from a disease known as rural poverty, a pathology that can easily be cured by an adequate insertion into the market. Cultivators in other parts of the world have shown this to be the case by knowing how to liberate themselves from an unbecoming condition by signing contracts with entrepreneurs, selling their labor force to large landholders, or shedding their peasant nature to become a part of nonagricultural economies. This resounding discourse creates the malady to legitimize the cure that institutions or corporations can offer. Perhaps it is true: this woman may today be in a crisis. But the diagnosis is wrong: she and her large family do not live in precarious conditions due to a lack of development, but rather because of an excess of development that has destroyed the means by which they might be able to get along in life and overcome their difficulties by using their own resources.

Back in the pre-development era, most people around the world could face difficulties, but it was outside their realm of ideas to conceive of poverty as an economic affair and, more remotely, that their situation might reflect personal shortcomings. Poverty, in the sense of an economic rationality of development, can only occur in an individualistic society, not where sociality is regulated by principles of reciprocity and mutual aid. In non-individualist societies (as a good proportion of traditional societies must have been before the expansion of capitalist modernity), a group was rich or poor in its entirety, and thus it was unimaginable that a family could go hungry while satiated neighbors looked on (Latouche, 2007). As Jean Robert and Majid Rahnema (2015) assure us, before the large-scale modernizing project, rural societies subsisted thanks to a network of concrete relationships that allowed them to find concrete solutions to concrete problems, facilitated by tacit agreements such as cooperation and reciprocity. Yet the dissociation of people from their autonomous habitability meant that they had to create linkages of dependency with the market and destroy the vernacular ways of living and thus lose the ability to act to solve the vicissitudes of life with their own means.

In terms of the rural producers of the Global South, first their traditional wisdom was destroyed by inserting them into the Green Revolution and its technologies. Once their ability to shape and maintain their livelihoods according to the ecological and cultural conditions of their inhabited place was dispossessed, they were forced to earn a living as agricultural day laborers, being unquestioningly tied to producing for extractive agribusiness, or migrating to the cities to obtain an income within the capitalist networks of exploitation. Ivan Illich calls this modernized poverty, i.e., a type of poverty typical of the developing world in which people, "expropriated of their possibilities to create relations among themselves and act jointly based on their own best interests" (Rahnema, 1996, p. 207), cannot find a way of subsisting other than living from day-to-day in the cities by doing work that is barely better than begging. Should they be lucky enough to find a paid job, they are forced to sell their labor at very low prices, making it impossible to satisfy their addictive needs created by the system itself (Robert & Rahnema, 2015).

As we have repeatedly seen, capitalist development does not use chains or compulsion. By touting a dream that democratic society is made up of free individuals and that the life of each one depends on individual effort and their ability to "get ahead," capitalism creates the expectation that people can be masters of their destiny (Roudinesco, 2000). Based on the aphorism "you too can participate," capitalism creates the belief that it is possible to choose the life that each individual wants (Bajoit, 2009). It is just a matter of becoming a part of the system so that we can make of our lives exactly what we always wanted. Nonetheless, the modern poor face constant frustration, since they are forced to live in a world plagued by products and services that inundate the market, but which are out of reach given that the poor have been dispossessed of the ability to "satisfy the desires that the system itself stimulates" (Castro-Gómez, 2015, p. 92). Given that they were led to believe that anyone can control his or her life and shape it according to personal desires, the frustration that arises when products are unobtainable is judged a personal failure, not the structural failure of the system (Marcuse, 1986). We can compare such a scenario to the Greek myth of Tantalus—as noted by Illich (2013)—in which the poor, like Zeus's son, are condemned to live in a paradise full of delicacies, but as they seek to quench their thirst or hunger, the fruit-laden branches move away.

Development skillfully creates expectations in the guise of promises. As Slavoj Žižek points out, the secret lies in promises never being fulfilled so that the collective force of continuous want is kept alive. Capitalism's discourse, like the one proclaimed by international organizations that orient agriculture on a global scale, dangles an object-desire before rural people, i.e., technologies that lessen agricultural workloads, inputs that raise productivity, calves born of cattle-fair champions, and contracts that guarantee income from harvests. These are permanent invitations to grasp emotionally at everything that is lacking. In Lacanian terms, subjects do not desire autonomously, but rather desire a desire created by others. It is a desire structured by institutions that places before the modernized poor an object of desire that permanently tempts them, but which, in the manner of a cat being teased, the amusement comes by constantly pulling the object-desire away from its reach. Thus, the poor's behavior is channeled through subtle symbolization that guides attention toward scientific-technical pleasures that are supplied at the market. The purpose in the end is both to have rural populations desiring modernized practices and to capture their desire such that their impulse to continue desiring is continuously energized (Žižek, 1992).

This way sensibilities are created that are shaped by a logic of interest (Machado, 2014), "I-ist" habits are evoked, affections are self-centered, and subjects yearning for objects and services that institutions (and more recently large agro-industrial corporations) can satisfy. Moreover, consensus is created regarding a belief that the modernized life provided by development will emancipate the rural poor from their unbecoming and shameful condition, which governments are guilty of by not implementing more programs and more projects that *insert* those *excluded* from economic growth. Rural development is an elegant and subtle force that bases all of its power on the construction of scarcity and shortage. To this end, it establishes an intricate system of needs (Illich, 1996) implemented through home censuses and surveys. Planners first establish what "should be," the correct and incorrect way of living, in order to put together a subtle series of indicators that are applied through prepared forms. Census takers visit homes and ask standard questions in order to transform interviewees into statistics and certify their poverty. Thus the construction of scarcity, previously created in government offices by bureaucrats or by development experts, is "scientifically" legitimized. Under the objective cloak of demographics, the need to intervene becomes clear in order to generate prosperity, encourage inclusive growth, develop new markets for environmental services, and stimulate territorial competitiveness by providing the poor with financial services and favoring large-scale private investment for job creation.

Once persons are convinced that they lack certain "indispensable" goods and services, they will not resist participating in development projects, since those same "participants" will themselves propose interventions that focus on speeding up the modernization of their lives (Rahnema, 1996). Indeed, not only will they not be an obstacle to the expanding interests of capital, they will be among its allies, likely demanding more chemical inputs, more hybrid seedlings, more commercial hens, more soft credits, more subsidies, more technical assistance, and more "support" for production. Greater power is granted when multinational mining and petroleum corporations undertake prior community consultations, bringing plans to compensate local residents and legitimize the large-scale operations that will be pursued in their territories. Under this scenario, local—coopted—populations will ask the corporations to build more roads, health centers, or schools as remuneration for letting them carry out their enormous investment projects. "Participatory" development is an imperceptible yet powerful means of manipulation that is useful in legitimizing "development" and promoting annual reports that claim that local residents chose to become part of the globalized economy through a "bottom-up" decision-making process. Mention should be made of the clientelism and loyalties that are created by this type of agreements that perpetuate the power of local political despots.

By stimulating the desire to be free and cut the ties of poverty, affections are created, and community affections are captured. An affective deterritorialization is implemented, redirecting desires and sensibilities toward the abstract time of progress. They also build deterritorialized imaginaries, as well as corporalities and sentiments that favor accumulation and consumption, separating bodies from territories, putting an end to affectations to the place and to neighbors, and estranging people from communal habitation and of feelings of belonging to the land. I insist that this process does not create insensibilities, but rather channels sensibilities, distinguishing that which can be felt from that which cannot be felt (León, 2011). By implementing a regime of sensitivity, bodies are separated ideologically from the inhabited place, distancing affections from spaces, re-signifying those same places, and inscribing them in development discourse. Thus, those same places are given a distinct meaning (Grosfoguel, 2016), desacralizing, materializing, and standardizing them according to the paradigm of competitive advantage. This much more discrete dispossession, i.e., a redirection of affection that tied many people to their living territories, reintegrates bodies into the desiring impulses of profit, increased productivity, and competitiveness.

Policies such as the ones we have discussed, which are poised to shape the future of agriculture by 2050—according to reports from international organizations—have a powerful affective weight that aim to economize the inhabitation of rural people and calm and appease political dissention. Expropriation practices, such as outright landgrabbing, are the most visible practices; it is easier to rebel against them. But the exercise of democratic policies of inclusion turns out to be a dangerous form of totalitarianism, as noted by Žižek, since it hides the power relationship and thus makes it more impenetrable. When development discourse surreptitiously says, "I decide what your need is, because I know better than you what you truly desire," it is really undertaking a profound colonization of the structures of desire. This is like a hidden voice that constantly repeats, "What you desire is to modernize your practices, be included in the market, and capitalize your life so that you will cease to feel like you were left out of development." This occupation of bodies, desires, and speech is so powerful that its articulation by the "target population" appears as a free and legitimate desire that emanates from the depths of one's soul. This is why it is so difficult to rebel, because once people's emotions, sensations, and desires are colonized, any criticism becomes lodged in the logic of development's economic rationality. A *habitus* is generated that becomes part of the lived experience and intimacy of the colonized.

As Spinoza (2011) said, an affectation on a body strengthens or weakens the power to act. With development, an affectation of desire increases the power of a type of action that resonates according to the logic of accumulation and the pace of progress. On the other hand, it decreases the power of action given the devastation of life and the misery of other human beings. In terms of the latter, the distribution of the sensible is more efficient when political response is prevented, as opposed to the regimes of desire that the system itself cultivates in the bodies of the colonized.

Esthetics of Agro-Industrial Progress

> To ask, "Where do you live?" is to ask in what place does your existence shape the world. Tell me how you live and I'll tell you who you are.
>
> Iván Illich

Another piece of the puzzle is still missing that we need to understand the power dispositives behind the expansion of extractive agribusiness. In this final section, I posit that the effectiveness of administrating and regulating bodies and the movement of sensibilities as discussed herein have much to do with the characteristics of the agrarian esthetics produced by capital. We use the word "esthetics" in its original definition derived from *aisthesis*, i.e., the intensity of perceptions from our senses, to highlight the relationship between the ability of our species to transform ecosystems and the perceptual experience of the world that emerges from this transformation. To cite Patricia Noguera (2012, p. 22), we assume that "the modifications that we as human beings make on ecosystems are modifications that we make on ourselves." Thus, agribusiness and its hierarchical-linear esthetic are inseparable from human perceptions and the affective meaning of people who live in close proximity to agroextractive geographies.

Let us revisit Tim Ingold's (2000) phenomenological argument to clear up my hypothesis. According to Ingold, we human beings arise in an environment, which means that there is no way to understand what is human without accounting for the specific settings in which people participate. The corporality mentioned previously as inherent to the human condition cannot really be considered independently from the communication that our bodies establish with the components of the inhabited environment. The corporality of our species, like that of other animals, is always in tune, resonating, vibrating, and connecting to the surrounding spaces in which we find ourselves. I wish to insist, following Ingold, that we cannot conceive of the environment as an exteriority nor can we set up artificial separations between our bodies and the space in which we live. We are a constituent part of an environment that we help shape through our actions and which affects us in our intervention. This is the reason why any affectation that we do to our surroundings is simultaneously an affection that we do unto ourselves.

Focusing on this process of mutual affectation, I wish to call attention to those agro-landscapes of development in which the straight line dominates, for example, furrows of soy that converge at a vanishing point; laying hens crowded in rows of cages; pigs confined to quadrangular pigsties that are divided by long rows; greenhouse domes organized into rectangular polygons; cattle stables arranged in geometrically standardized chutes; and palms that stretch monotonously in endless rows. These are examples of the planned landscapes in which technology and architecture have imposed particular norms on ecosystems in an effort to regulate the lives of the humans who live therein in accordance to the criteria of efficiency and maximum profitability (Lefevbre, 2013). Although it is true that "we are what we do," following Maturana and Varela (2003), these agro-industrial landscape creations are simultaneously and continuously creations of those who undertake these modifications. If phenomenology is correct in affirming that there is no irreducible separation between body and space, then this type of landscape esthetics is inseparable from human corporalities and sensibilities, since both environment and bodies are bound together in a permanent, reciprocal relationship.

I propose enquiring to what extent these ecosystemic transformations have become embodied in our intimacy, creating competitive, isolated, solitary, and distrustful beings. And if in some way these spaces, modified by metaphysical economic rationality, have fully inhabited the bodies of those who participate in these landscapes. Places inhabit us way before we inhabit them, according to José Luis Pardo (1991). Thus, it is not absurd to ask how mutilated and flattened lands that have been disciplined to become pasturelands and fields of monocrops, as well as the architectural interventions that are essential features of agro-industrial esthetics, are embodied in those who live in proximity to those places. The Green Revolution esthetic is in my opinion the quintessential metaphor of the image we have of ourselves and a radical expression of the ontological conception of our modernity.

Therefore, we are not only devastating the Earth. During the geometric ordering of spaces suited for economic accumulation, varieties of human "beings" coherent with such a violent transformation are also being created.

Recall Heidegger's argument that modernity is an era in which the world appears as an ordered image. For Heidegger (1996), modernity is an era in which we human beings categorize entities as objects while simultaneously granting ourselves a privileged position as subjects. According to this German philosopher, solely to the extent that we occupy such a position is it possible for the world to become an orderly world-object, calculated and managed according to logical-linear-teleological thinking characteristic of modernity. Clearly, the previously mentioned esthetic registries are a radical expression of this dominating position of modern culture in which ecosystems are occupied, intervened, manipulated, transformed, and ordered into an image-object. Yet human bodies also inhabit this new world, and thus our emerging behavior will depend on human perspectives in a setting that has been transformed into geometric figures that are counterposed to the organization of life. It is in the heart of these esthetics in which the perceptible regime arises in one way and not another. Therefore, according to how the world-object is molded will depend on how *ordo amoris* of the people living in these contexts is shaped.

Perception changes as the inhabited world varies. A particular perception takes shape when the feeling of pertaining to the earth is hollowed out and then replaced with an attachment to desire, stimulated by a system where subjects actively participate. In other words, the production of spaces that has disciplined plants and animals, imposing rectilinear shapes onto the landscape (Lefevbre, 2013), has also created a functional perception in agricultural producers and inhabitants who live nearby to those modified places. This is a type of perception of the environment that is profoundly associated with the ways of "being" in those places that have become monotonous agro-industrial landscapes. If our encounter with nature occurs in a linear, delimited, standardized, controllable, and profane environment—such as the immanent esthetics of agroextractivism—then to what extent and under what conditions is the perception of the surroundings of communities in which such phenomena occur co-created as a construction of representations, meanings, and feelings that are apropos to the mechanical assemblage of industrial crop fields.

My hypothesis, which still needs fleshing out with empirical evidence, is that the affective tonalities of rural societies, the regulation of perceptions, and the perceptible ordering of the population that lives in proximity to those places are associated with the production of spaces that are coherent with the factory model as mirrored onto nature (Shiva, 2007). This occurs because the geometric organization of rural fields adapts to and is relatively aligned with daily existence. The perception of agricultural producers, like that of their fellow human beings, depends on the environments in which they participate. Thus, we should ask how the world would be seen within a linear, subjugated, and disciplined environment, such as the immanent esthetics of extractive agribusiness.

In the end, I believe that the industrial expansion of agribusiness involves a fundamental ontological transformation for people who are a part of these spaces, transformed into precise, exact, and calculable figures of Euclidian geometry. I refer to the many communities that have resisted selling their lands and thus are trapped between enormous plantations or are in situations where residents have no other choice than to work in the agro-industrial enclaves as agricultural proletariats. A third group of agricultural producers has turned to agricultural corporations that produce mainly for the export market. All these corporalities are forced to live amid the tensions of a major landscape change. For these reasons, I believe we need to lend greater attention to the perceptive transformations that arise from the agroextractive mutations that have occurred mostly since the start of this millennium.

One example is worth mentioning. The so-called soy republic is a land surface that has grown from 17 million hectares to 46 million hectares between 1990 and 2010. It covers extensive areas of Brazil, Argentina, Uruguay, Paraguay, and Bolivia where some 20 million hectares have been deforested between 2000 and 2010 (WWF, 2014). This is an obscene ecocide led by agricultural corporations and backed by the governments of these countries. This is also, I stress, an affective and perceptive dispossession, if such a term can designate the biological power that colonizes the most intimate sphere of sensibilities by means of the esthetics of agro-industrial progress. As biological beings, we are always situated entities; thus we must focus on the experience of a world that emerges in environmental contexts, such as these geographies. The affective tonalities of economic rationality inevitably emerge against an occupied, controlled, and technologically regulated backdrop as stipulated by agroextractive esthetics. This is a scenario where daily experience transpires, affections are regulated, and time is administered, while behavior, desire, wisdom, and truth regimens are produced.

This is obviously an undetectable exercise of power, but no less powerful. This process codifies, inscribes, and registers affective flows in the geography of agribusiness and, in general, in the esthetics of development. In sum, I believe that effective territorial control cannot exist without the colonization of sensitive structures, moods, feelings, and desires of the hegemonized. This sensitive regime is based on the production of ordered spaces in keeping with the industrial esthetics of the assembly line.

In these two chapters, I hope to have shown that the study of agro-capitalism's ceaseless expansion can be more rewarding when we analyze the domination of territory for the purpose of appropriating rents. This is more properly done by including populations within the system, as I have discussed based on the idea of the ontological and epistemic deterritorialization of rural peoples, aided by the megamachine of development and the production and use of subjectivities. I do not want to deny that expropriation, in the manner of Marx's primitive accumulation of capital, has an important place. In fact, as we shall see in the following chapter, because of the crude exercise of power during the current neoliberal period, political opposition has grown. I have not given, however, as much attention to episodes of dispossession as I have to the more elegant exercises of power, as described by Gronemeyer (1996, p. 8):

> The defining characteristic of elegant power is that it is unrecognizable, concealed, supremely inconspicuous. Power is truly elegant when, captivated by the delusion of freedom, those subject to it stubbornly deny its existence... It is a means of keeping the bit in the mouths of subordinates without letting them feel the power that is guiding them.

For a period we cannot yet define, development has been and will continue to be the most powerful mechanism of splitting people away from their material and symbolic conditions of existence and the most sophisticated device for furthering the geographic growth of capitalism in the countries of the Global South. Development undertaken as accumulation by dispossession is the best way I have of defining the way that agricultural extractivism continues to territorialize itself in the crop fields of Earth's surface. It is up to those of us who fiercely oppose this violent expansion to reject not just its most apparent mechanisms of dispossession but also its least visible ones, i.e., those we cannot detect because to some extent we have taken a plunge into their interior.

Chapter 5
Agroecology in Post-development

The problem with ballooning globalization is that balloons can burst.
Nabucodonosor, Escarabajo

Up to this point, we have viewed capitalist modernity as if it possessed an almost magical power to invade and corrupt everything, or the omnipresent ability to colonize every corner of the planet, or had the power to structure the world according to its whims, transforming persons and ecosystems according to the laws of the market. While this might be the desire of capitalism's proponents, it is unlikely always to achieve its end. In spite of the drive to bring all social relations into global value-added circuits, the other side of the equation consists of struggles, resistance, and hidden power strategies that are outside capital's field of vision (Gibson-Graham, 1997; Scott, 2000). People are not passive agents nor puppets that move according to the whims of the power that engulf them. Undoubtedly, power has the ability to overwhelm, but subservience is never total. Further, when power loses its sophistication and its prosaic and violent nature is exposed, it produces an antagonistic effect by mobilizing resistance. This is precisely what has occurred with the consolidation of capitalism's neoliberal phase throughout the world.

Ecology became political because capital's growing interest in appropriating nature generates antagonism, conflict, and battlegrounds for the defense of territory and life. Social movements throughout the Global South take up the cause to oppose accumulation by dispossession (Harvey, 2003). These struggles that resound throughout rural and urban grassroots settings are clashing with capital over water, land, seeds, and knowledge. Agroecology has informed these battles and occupied the arena in order to bind together grassroots efforts to address the unjust distribution of wealth, environmental pillaging, unsafe foods, hunger, malnutrition, country-to-city migration, and the growing proletarianization of rural inhabitants due to the current

O. F. Giraldo, *Political Ecology of Agriculture*,
https://doi.org/10.1007/978-3-030-11824-2_5

market-based agricultural system (Rosset & Martínez-Torres, 2012).[1] Specifically, as Peter Rosset (2016, p. 00) writes, "rural social movements, made up of families from peasant, indigenous, or other rural populations, are actively defending rural spaces, competing with national and multinational agribusinesses, as well as with other private-sector actors and their allies in government." The political ecology of agriculture is indeed the battleground where social movements are struggling over agricultural hegemony. The battle is not just over its political instruments but particularly the ontological, epistemic, and ethical conditions that allow for the construction of territorialities composed of diversified croplands, community forests, mountains, and rivers—which ultimately is a compelling image from their rural utopia—in opposition to large landholding agribusinesses with their green deserts bereft of small-producer families (LVC, 2015b).

This utopic imaginary of the social movements is what mobilizes people into political action to confront the current hegemonic system. This is not about peacefully coexisting, where large- and midsized agro-industrial properties, cattle haciendas, and monocrop extensions can sit alongside small rural producers and agroecological communities located on the least fertile lands. Small producers are waging an ongoing agrarian struggle for a total redistribution of land and a reconfiguration of the overarching agricultural and food systems where the agroextractivist and large landholding structures would be banned. Besides being a protest against capitalism, this refusal to coexist is also grounded in ecology, given that agroecological science has shown that a diversity of plants and animals in the fields, forests, and wilderness areas can help boost biological control, pollination, and soil fertility. In contrast, ecosystemic simplification and the use of agribusiness's agrotoxins in neighboring fields could inhibit this type of ecological activity over the entire landscape (Vandermeer & Perfecto, 2010). This is why social movements' struggle—such as La Vía Campesina's—over land and agroecology is both a dispute against monopolistic tendencies in the means of production and a battle for control of agriculture's technical characteristics.

In opposition to the mainstream model, the utopia of small producers, indigenous, fisherfolk, nomadic, and other grassroots urban and rural sectors is about creating

[1] LVC (La Vía Campesina) (2015a, 2015b) declared at the *International Forum for Agroecology*, "Agroecology is the answer to how to transform and repair our material reality in a food system and rural world that has been devastated by industrial food production and its so-called Green and Blue Revolutions. We see Agroecology as a key form of resistance to an economic system that puts profit before life. The corporate model over-produces food that poisons us, destroys soil fertility, is responsible for the deforestation of rural areas, the contamination of water and the acidification of oceans and killing of fisheries. Essential natural resources have been commodified, and rising production costs are driving us off the land. Farmers' seeds are being stolen and sold back to us at exorbitant prices, bred as varieties that depend on costly, contaminating agrochemicals. The industrial food system is a key driver of the multiple crises of climate, food, environmental, public health, and others. Free trade and corporate investment agreements, Investor-State Dispute Settlement agreements, and false solutions such as carbon markets, and the growing financialization of land and food, etc., all further aggravate these crises. Agroecology within a Food Sovereignty framework offers us a collective path forward from these crises."

agroecological territories where healthy foods and nonfood items are produced, in tune with nature's cycles, using grassroots knowledge, and where young people and other family members choose to live in the countryside (Rosset, 2016). This praiseworthy goal requires confronting the power of capital, one of the most important aspects of which is the concentration of land. As Hegel teaches in the dialectic of the master and the slave, land rights are nothing more than a right recognized in notarized and state-recognized land titles that lose their value when subordinates disavow the documents. Recall that the master is not the master just because he "is," but because the slave recognizes him as the master, just as the king is not king because of some divine blessing, but because his subjects recognize him as king. Similarly, public titles and the enclosure of property that ensures monopolistic control of the land are not legitimate due to the grace of God, but because the landless recognize (through state mechanisms) this type of property as a legal and binding right that must be honored. Land occupations and settlements under agrarian reform, such as those implemented over the years by the Landless Rural Workers Movement (MST) in Brazil or the Small Organic Producers Forum in Zimbabwe (ZIMSOFF), are a direct affront to this recognition. In essence, these organizations have ceased to recognize the bourgeoisie as a rightful owner, and by tearing up their land titles or pulling down their illegal enclosures, they subvert the existing order, much as when slaves and subjects stop recognizing the master and the king as "master" and "king."

It remains to be seen whether water shortages and contamination, the effects of climate change, growing ecocide due to the voracity of capitalist accumulation and, thus, increasing territorial control, will finally erode the fragile recognition that the dispossessed continue having for large rural landholders and thus become a massive movement. This might happen not so much as a revolution against the state as administrator of the dominant classes' interests—as classical Marxism predicted—but rather because of a pragmatic response the underclasses' to guarantee their own survival. This is what is at stake given the environmental crisis and current general trend of dispossession (Giraldo, 2014). We have previously mentioned that capitalism is unable to refrain from monopolizing land and destroying its production conditions. Indeed, this voracity has led to a collective exasperation, social organization, and the announcement by environmental activists around the world that they will challenge capital's current "project of death" and propose a shift to a "project of life." As Pierre Madelin (2016) has stated, these activists are both questioning the socioenvironmental effects of the capitalist way of production and proposing alternatives such as agroecology; they are also undertaking a political critique of monopolies and struggling against heteronomous technological structures that can only be managed by a centralized power.

In this chapter, we analyze the agroecological political alternative proposed by social movements that is not simply a modification of institutional practices but also a different way of practicing the art of politics (Lefort, 2004). It is based on radically different principles that inform us as we seek to transition to post-development and, more importantly, to a postcapitalist world.

Social Processes of Agroecology

The great modernizing development project begun in the mid-twentieth century has been an apocalyptic nightmare. There is general agreement that we are on the cusp of a new geological era, called the Anthropocene—more aptly called the Capitalocene (Moore, 2017)—that coincides with the start of this mega-civilizing project (Waters et al., 2016). Just in terms of agroextractivism, we have seen not only how the development project has disrupted the planet's ecological systems but also how it has sparked a profound multidimensional crisis that has created serious danger and iniquitous distributive injustice. The emergence of neoliberalism deepened the capitalization of nature, heightened policies leading to dispossession, and (following the financial crisis) encouraged a new extractivist tendency to privatize, control, extract, and commercialize biodiversity, water, land, minerals, and hydrocarbons to be exported as commodities on the world market (Composto & Navarro, 2014). In agriculture, extractivism includes the classical rent-based monoculture, fishery and forest extraction, large-scale cattle ranching, and agrofuels, which have created numerous new territorial conflicts in countries of the South.

This new extractivist tsunami of accumulation by dispossession heightened global capitalism's tendency to deterritorialize, tearing down rural economies while increasing national and international migratory flows. Yet the voracity of capital accumulation has also led people to strengthen and defend the commons in light of the immense capital onslaught, whose intention is to bring all remaining territory into the sphere of market valorization (Navarro, 2015). Many communities have organized and have called for a defense of their territories—leading at times to a loss of life among their members[2]—as a resistance movement in the face of capitalism's interest in controlling everything it can. It is within these organizations where agroecology and food sovereignty, or, better still, food sovereignty through agroecology, have become guiding principles for the struggle with capital over territory, in an attempt to question the material conditions of production and, as Polanyi might say, to embed technology and the economy with the social.

Due to the system's own contradictions, grassroots organization has grown, encouraged at times by a collective weariness of the grim consequences of the agro-industrial model that informs the Green Revolution. I refer to the environmental effects of the spread of monocultures, the increase in pests and diseases previously unseen in croplands, the loss of soil fertility, the water contamination, and the growing vulnerability of agroecosystems to natural disasters. The expanding grassroots organization is also a rejection of the vulnerability created by this system to macroeconomic fluctuations, such as devaluation, price increases of agricultural inputs, and the drop in prices for agricultural products. And also due to people's anger at the deterioration of their health due to agrochemical toxicity and their chronic illnesses

[2] According to the NGO Global Witness (2015), every week two environmental activists are murdered somewhere in the world.

brought on by a highly industrialized diet. These concerns have been addressed by large social movements, such as the MST, that in 2002 integrated agroecology into its program of action for the 350,000 families who occupied and recovered land. It is also the example of La Vía Campesina International, whose member organizations represent more than 200 million families on 5 continents (Rosset & Martínez-Torres, 2012). Yet more often, the search for alternatives is not the result of direct collective rejection of development or a direct political rebuff of capitalist agribusiness. Rather, this is often a self-protection and not always spontaneous mechanism, frequently encouraged by external allies, such as Catholic and Protestant churches, and NGOs that use Freire's popular-education methodologies.

Regarding this latter aspect, someday there will be recognition of the largely untold, revolutionary role played by liberation theology in the spread of agroecology throughout Latin America. I would go so far as to claim that since the mid-1960s, rebel parish priests, nuns, and laypersons silently created a revolution in rural areas that was not only barely acknowledged but also made invisible due to the left's insistence on taking over state institutions. In this regard, I do not mean to belittle the fact that, with exception of Cuba, in recent times the inequities of land distribution have become more pronounced,[3] and Latin America again claimed a dishonorable first place in being the most unequal region on Earth. I do wish to underscore the very important role of liberation theology and, especially, Indian theology in enhancing traditional worldviews and the knowledge that was lost due to development policies or because of market influences. Learning methodologies were implemented through the "dialog of wisdoms" and action/reflection/action, as a way of self-managing one's affairs to vindicate community spheres that had eroded in the face of advancing globalized capitalism. In many different ways, liberation theology has been the foundation for various concrete, small, discrete, and very dispersed experiments that were based on spirituality[4], i.e., not on economic rationality. This spiritual undergirding made them especially resistant to attempts to co-opt and "capture" them. Generally linked to pastoral ecclesial movements, NGOs have been a key element in encouraging numerous agroecological initiatives that have discretely rewoven ecosystemic relations and regenerated social linkages.

[3]See the data analysis of Robles and Concheiro (2014) regarding land concentration in Latin America.

[4]At a 2010 conference at the College of Economics at the National Autonomous University of Mexico, Michael Löwy made a provocative association between Latin American liberation theology and Benjamin's (2008) first thesis on history. In this famous and celebrated text, Benjamin uses the metaphor of a chess-playing robot that wins all matches. In reality, a master chess-playing dwarf was hiding inside the robot. For Benjamin, the robot represented historical materialism and the dwarf theology. With this analogy, Benjamin proposed that Marxism uses theology for its purposes. At the conference, however, Löwy held that in Latin America liberation theology had been the hidden dwarf of a less ostentatious revolution yet persistent enough to change the lives of hundreds of thousands of the dispossessed.

It is difficult to keep tract of these small but vivid experiences where liberation theology has played a role. It was especially the Jesuits, but also the Dominicans, who were key players in promoting liberation theology. It is perhaps the fact that these experiences were not ostentatious, but rather persistent and, perhaps without recognizing it, in terms of their social relations, they have traveled down a very different road than the one taken by development and its industrial tools for capitalist expansion. Although they were not as political as some hoped, these widely disseminated efforts in nonurban areas are examples of how to build something entirely different from neoliberal capitalism.

Throughout an untold number of peasant, indigenous, and Afro-descendant communities in Latin America and the Caribbean, for decades liberation theology and lay support organizations have promoted agroecology (although not always with that name), through methods based on a dialog of wisdoms and social creativity. At times, these methodologies contain certain messianic principles, such as the need for critical external intervention or the need for mediators as agents of liberation of the oppressed (Esteva, Prakash, & Stuchul, 2002). Yet undeniably, liberation theology has been a seed that germinated in other forms of learning, such as, in my opinion, the most important contribution to post-development, i.e., the peasant-to-peasant methodology.

According to Eric Holt-Gimenez's book, *De Campesino a Campesino: Voces de Latinoamérica* (Holt-Gimenez, 2008), the history of this methodology, which is worth analyzing in detail, began in 1972 in Chimaltenango, Guatemala. The process started after agronomist Marcos Orozco had unsuccessfully tried to promote containing walls, terracing, and organic fertilizers among Kaqchikel indigenous peasants. The peasants were neighbors of Orozco's and had suffered serious problems of erosion, productivity, and indebtedness years after adopting Green Revolution technologies. The NGO World Neighbors, emulating the Guatemalan Barefoot Doctors, offered to train some of the Spanish-speaking peasants known to Orozco as promoters of his methods in indigenous communities. With great difficulty at first due to peasants' mistrust of suggestions made by specialists, soon the small producers agreed to implement small trials on their plots of land. Once they had seen the efficiency of the suggested practices on a small scale, the techniques were implemented on their entire fields, and promoters began to share with neighbors previous forgotten methods that had been all but forgotten. Thanks to mutual aid and community work, the fields of the rural producers who worked with Orozco began to change and to have stunning results. Over time, they became promoters of agroecological and increasingly more complex agroecological practices, sharing their knowledge through workshops, field visits, and agricultural fairs, spreading their knowledge throughout the region. Oxfam and World Neighbors accompanied the founding of the Kato-Ki cooperative, which at one time had almost 900 members, and facilitated several exchanges with peasants from other regions of Guatemala and Mesoamerica (Holt-Gimenez, 2008).

The horizontal socialization of agroecology grew surprising quickly during the early 1980s. Yet, precisely because of its notable achievements, the Kaqchikel peasants faced enormous pushback from local large landowners. The peasants' activity both reduced the amount of labor available for the coffee plantations and made them direct competitors. The regional bourgeoisie accused them of being communists, and backed by this McCarthy-era accusation, the Guatemalan army carried out brutal persecution, forcing them to flee and abandon their lands, and eventually to dissolve the Kato-Ki cooperative. With the support of some NGOs, the promoters sought refuge in Mexico, Honduras, and Nicaragua, where they continued their role as promoters by encouraging agroecological practices. Specifically, the Vicente Guerrero *ejido* in Tlaxcala, Mexico, is where they continued their practices. It was there where the local small producers, who years before had visited a rural agroecological experience in Chimaltenango and had implemented soil conservation practices, welcomed two of the promoters who had fled their homeland. Without external intervention, the new promoters were able to consolidate the dissemination of agroecological practices directly to local inhabitants (Holt-Gimenez, 2008). Although the results on the ground materialized rapidly and are still visible today—the Vicente Guerrero experience continues to be an important point of reference in Mexico—it was in Sandinista Nicaragua where the methodology was transformed from a local occurrence to a mass phenomenon.

In 1986, a group of promoters from Vicente Guerrero, with support from the NGO Sedepac, was invited to Nicaragua by the National Union of Agricultural Producers and Cattle Breeders (UNAG), a rural association created following the revolution and implementation of agrarian reform policies. After several meetings between agricultural producers, workshops, and round-trips between Mexico and Nicaragua, results were soon forthcoming. Nicaragua was waging a war, facing an economic blockade by the United States, experiencing unprecedented hyperinflation, and weathering a serious food shortage with seriously eroded croplands (Vásquez & Rivas, 2006). It was in Santa Lucía where the fruits of the exchanges were first seen. In just 5 years, eight promoters had greatly expanded agroecological practices in the region. Santa Lucía became a beacon of inspiration for the creation of the Peasant-to-Peasant movement throughout the country (Holt-Gimenez, 2008). In 25 years, the network of promoters grew from 11 peasants in 1987 to 1918 in 2012, who exchanged knowledge with 15,000 rural producing families throughout Nicaragua (Vásquez & Rivas, 2006). There are estimates that peasant-to-peasant has been able to promote agroecological practices among 35,000 families. This exponential growth is largely because agroecology was embraced by a national organization such as UNAG, as opposed to Honduras, Guatemala, and Mexico, where it was not taken up by national producers' organizations. Much of the success, however, has to do with the multiplying capacity of methodologies based on the local promoter (Machín, Roque, Ávila, & Rosset, 2010).

The "promoter" is the agricultural producer who successfully applies some technology on his/her own plot of land and later promotes or encourages other producers to adopt it on their plots. Promoters promote dissemination of agroecological knowledge with their neighbors, since with their example they incentivize local producers to experiment and innovate creatively. The process begins when producers from different communities visit promoters who successfully used a technique on their own plot. Upon returning to their plots, they experiment or implement some aspect that interested them (Vásquez & Rivas, 2006). Thus they slowly create new knowledge that is exchanged with other families, increasing the transformation of landholdings that initially depended on external inputs derived from fossil fuels and planted as monoculture crops but now are relatively autonomous and diversified landholdings, based on local innovation and the use of solar energy. The major difference between this methodology, compared to classical development extension programs, is that, on the one hand, what is disseminated is not based on fixed recipes, but rather flexible principles that can respond and adapt to every particular situation. Further, peasant and indigenous communities are the protagonists of the entire process. Thus creativity is encouraged and so too the ability of rural communities to experiment, innovate, evaluate, and widen their store of knowledge and innovative skills in order to find solutions to their own problems, rather than passively waiting for experts' recipes (Altieri & Toledo, 2011). The effectiveness of the methodology also lies in that peasants put more faith in what another rural producer does than what an extension agent says and, thus, is more likely to emulate a practice seen when visiting another peasant's land and physically seeing the results (Machín et al., 2010).

The Vicente Guerrero group and the UNAG shared with Cuba an impressive and invigorating methodology, which promoted autonomy during this island's profound crisis known as the "special period in times of peace." Cuba had specialized in producing sugarcane, tobacco, and coffee, becoming dependent on imports of petroleum, subsidized foods, machinery, and inputs such as fertilizers, pesticides, and herbicides from the Soviet bloc. When the Soviet Union fell, Cuba suffered a severe food crisis, due to its limitations in importing food and necessary materials for agroindustry, the worsening US economic blockade, and the end of the agro-extraction model[5] (Machín et al., 2010). Need drove Cuba to become food self-sufficient with agroecological techniques in rural and urban areas, because it was impossible for the country to continue relying on an agriculture that depended on inputs and machinery associated with the Green Revolution. It was in this context that in 1997 a Bread for the World project to promote agroecology began in Villa Clara Province and then spread in 1999 to the provinces of Sancti Spíritus and Cienfuegos, thanks to support from Oxfam and the Catholic Committee against Hunger and for Development (CCFD). The turning point for its exponential growth, however, was 2001 when the ANAP (National Association for Small Producers) turned peasant-to-peasant into a national movement. As opposed to the UNAG, the ANAP took on agroecology as an integral activity within the organization's structure. Thus, the movement became more systematic and less spontaneous than previous efforts. Through a well-planned organization among cooperatives, facilitators, coordina-

[5]The following estimates were made after three decades of the Green Revolution: "43.3% of the soil suffers from erosion and 23.9% from compaction, 14.1% has high salinity and 24.8% is acidic; 44.8% has low fertility… Cuba used to import 48% of its fertilizers and 82% of its pesticides… (and) food imports made up 57% of total family calorie intake (Machín et al., 2010, pp. 42–44)."

tors, universities, and some government bureaus that were involved in granting access to land to a larger number of small producers, the critical period of food shortage was overcome. The results were spectacular. From 114 promoters and 5800 families in 2001, the movement grew to 11,935 promoters and 110,000 families in 2009 (Machín et al., 2010). Today, ANAP estimates that 130,000 small-producer families farm using agroecological methods, in addition to 530,000 urban patios that also use these practices.

Although Cuba's modernization has been amazing—it is said that of all the Latin American countries, this island nation was the most heavily involved with Green Revolution technology—ancestral knowledge still lingered among small producers. This knowledge became the foundation for techniques that initially substituted imports and then slowly incorporated agroecological techniques that solved structural problems in the agroecosystem, such as the lack of agrobiodiversity and ecological interactions on an ecosystemic scale.[6] Other important factors were a highly politicized ideological discourse against the agro-industrial system linked to the Green Revolution, the socialist values of the Cuban Revolution, José Marti's liberation philosophy, the pride of being small producers, food sovereignty, and stewardship of nature (Mier y Terán Giménez Cacho et al., 2018). Currently the methodology is bearing fruit in other areas of the world, such as Northern Mozambique with the participation of 4500 small-producer families (LVC, 2015b) and in Malawi where another 6000 families have joined their ranks (Kangmennaang et al., 2017). The movement has inspired the founding of 65 agroecological training schools in Venezuela, Paraguay, Brazil, Nicaragua, Colombia, Chile, Haiti, Indonesia, Thailand, India, South Korea, Spain, Zimbabwe, Nigeria, Mali, and other countries in the 5 continents (LVC, 2017).

The process that began in Chimaltenango also influenced practices in Southern Mexico. In 1989, given the prevalent neoliberal context, the Mexican Coffee Institute (INMECAFÉ) was dismantled, which coincided with other negative events, such as the collapse of international coffee prices, the rising price of inputs, and shrinking returns (Hernández-Castillo & Nigh, 1998). With the accompaniment of Indian theology within the Catholic Church, the opening of organic coffee and fair-trade markets and the exchanges between small producers that arose in a more spontaneous way—in part due to the presence of Guatemalan agroecology promoters who were displaced by repression in the 1980s—wisdom from the Mayan worldview was recovered, crops were diversified, and a mayor network of cooperatives was formed that turned Chiapas into the number one exporter of organic coffee in the world. While not as vigorous as the movement in Cuba and although questions arise because of its dependency on export markets and foreign certification require-

[6] Disseminated agroecological practices have included use of crop residues, green fertilizers, organic material, contour curves, crop rotation and association, biological control of pests, introduction of animals, increase of diversity, and use of alternative energy sources, such as the hydraulic ram, animal traction, windmills, and biodigesters (Machín et al., 2010).

ments that lessen autonomy,[7] growing coffee is currently the livelihood of some 31,000 small-producer families, mostly Tzeltal and Tzotzil indigenous peoples. In terms of post-development, a much more interesting agroecological experience in the same region is that of the Zapatista Army of National Liberation (EZLN). Guided by their principle of self-government and anti-capitalist autonomy, a network of agroecological promoters was established that lends increasingly greater food autonomy and economic self-sufficiency to the EZLN's grassroots supporters. While the movement does not divulge membership data, thousands of rebel families practice agroecology disseminated by the peasant-to-peasant methodology.

Although not related to the Latin American Peasant-to-Peasant movement, the Zero Budget Spiritual Farming effort in India[8] has interesting similarities. As in other cases, the movement emerged in a context of severe crisis provoked by the adoption of monoculture farming with Monsanto's GMO seeds, the indebtedness of farmers, and the effects of having implemented Green Revolution technologies. The crisis led to an unprecedented wave of small-producer suicides attributed to their unpayable debts. There are estimates that 290,000 small farmers have taken their lives since 1995. This perverse situation laid the foundation for a movement encouraged by Subhash Palekar, an agronomist who since the 1990s has undertaken trials and documented successful agroecological practices in his own plot. In 2002, Palekar, already well-known, was invited to the southern state of Karnataka by the Karnataka Rajya Raitha Sangha (KRRS) peasant organization to lead a 5-day workshop in a training field. Five thousand farmers attended. The natural agricultural techniques promoted by Palekar mainly consisted of bio-fertilizers, vegetation cover, integration of zebu cattle, local earthworms, ecological pest management, and plant association. These techniques were combined with a discourse that blends a critique of what Palekar calls an exploitive, anti-peasant system, controlled by multinational corporations and Western culture with Hindu mythology, holiness of cows, and Gandhian principles of personal change, simple living, non-violence, autonomy, and stewardship of Mother Earth (Khadse, Rosset, Morales, & Ferguson, 2017).

The training camps turned out to be instructive, but the most interesting aspect occurred when the peasants returned to their plots and tested the efficiency of the practices, which quite obviously did not work like recipes. Each farmer adapted them to his/her plot, which stimulated creativity and innovative capacity. Soon people began to volunteer spontaneously. These volunteers, similar to the Latin American promoters, disseminated their experiences through a methodology that is very simi-

[7] In contrast, other experiences, such as Ecovida in Southern Brazil, have adopted agroecology thanks to the demand of 200 local markets and their own self-managed participative certification. Estimates show that some 3500 agroecological families participate in this growing network—Information supplied by Professor Julián Pérez during the April 2017 "Permanent Seminar on the Massification of Agroecology."

[8] Information regarding this case in India is taken from Ashlesha Khadse's master's degree thesis at El Colegio de la Frontera Sur, "Movimientos Campesinos y Escalamiento de la Agroecología," [Peasant Movements and Scaling-Up Agroecology], published in part in Khadse et al. (2017).

lar to that of peasant-to-peasant. Once they successfully implemented these alternatives in their own plots, the farmers began to invite others to see their successes. The model varies and depends on local leaders, but neighbors frequently gather in a farmer's homes to analyze their plots and work plans and exchange experiences. They also hold workshops for other persons. At times, they travel several kilometers to follow up with newcomers, and they maintain regular phone contact. Thanks to some 60 training camps, the movement grew throughout the country and reached 4 million people according to the most optimistic estimates. In contrast to the Latin American cases, no formal structure organizes the tasks. Yet there are alliances with peasant organizations, NGOs, individuals within the power structure, *mathas*—religious institutions—scientists, writers, and non-mainstream media such as the Internet, community radio stations, television, social networks, and even WhatsApp applications (Khadse et al., 2017).

These movements that have grown exponentially in different parts of the world demonstrate the importance of social organization. In none of the cases were the rural producers isolated. An organizational structure was necessary for the process to grow. This organicity allowed the movements to challenge the hegemony of the mainstream agro-industrial food system. By *teaching through example*, they have shown other subaltern social sectors that agroecology has the ability to erode a set of supposedly unquestionable beliefs about agriculture propagated by the dominant class. For this to occur, scientists who have accompanied the agroecological movements with their research and other external allies have done important work. Nonetheless, the shared aspect within agroecological territorialization has been that producer organizations have taken charge of the process without depending on outside actors. Other key factors include the fact that the system suffered a severe crisis, thus stimulating a search for alternatives, there was access to land, effective agroecological practices, and a mobilizing discourse that defines agroecology as the preferred roadmap within distinct cultures (Mier y Terán Giménez Cacho et al., 2018).

The horizontal exchange of wisdom and specifically the peasant-to-peasant methodology is in my opinion agroecology's principal contribution to a post-development transition (Escobar, 2015). Indeed, I believe that practices undertaken by social movements are the most interesting attempts to subvert the logic of capital, as we discuss in the following section.

Regeneration of Community Spaces and Socialization of Vernacular Wisdom

The concept of post-development has been used in some academic circles to foresee the demise of "development" discourses and practices that have played such an important role in the expansion of agro-capitalism throughout the world. In Latin America, the notion of good living—*Buen Vivir*—has been the most important unifying concept for imagining post-development from the standpoint of social movements (Giraldo, 2014), similar to the concept of degrowth on the European continent. With this neologism, social movements hope to imagine a world where power relationships linked to large-scale development projects and imaginaries of growth, progress, industrialization, and modernity shift to a "pluriverse" through epistemic decolonization, communality, relationality, autonomy, the deconstruction of patriarchy, and post-extractivism (Escobar, 2015). This will ultimately mean that many diverging paths of capital will be built from below in harmony with the land.

This undertaking is not free from contradictions. At a more frequent rate than its defenders like to admit, the concept tends to be bandied about by theoreticians in a way that is useful solely to elite academic journals. Or, worse, in institutional settings the concept is revisited in terms and practices that are very similar to the "great development project" carried out over the past 60 years. Development has a very powerful semantic weight and has irradiated widely to all areas of knowledge, with its peculiar way of seeing the world. It has made thinking about truly alternative and viable practices very difficult. My hypothesis posits that the social processes described herein are precisely the space within which we can challenge development while simultaneously and dialectically imagining post-development in a pragmatic way.

To define the best road to take, I believe it is useful to begin with the questions raised by social movements that criticize the model in existence since Truman's speech. This is done in an effort to put aside the prevalence of its conditions, rules, and statements and seek signs about which of the alternative paths we should head down. The exercise inspired by Illich (2006b) is as follows: We first state what it is that we reject and then define, almost in opposition, what we would like to see. We have discussed, for example, how development increases dependency and the margin of control exercised by external institutions that manage communities' time and activities. Therefore, starting with the critique, we can deduce that the principles of post-development need to allow for the exercise of more autonomy, such that people can again take control of their immediate problems. We have also stated that development decides on people's needs, produces uniformity, and strips the individual of his/her creative ability. Thus, an alternative path would have to encourage people to outline, control, and limit the problems they face and their solutions, promoting creativity, social invention, and the collective imagination. It has also been suggested that development makes people vulnerable to expert knowledge and targets them for interventions. Therefore, ideal tools would need to encourage the recovery and exchange of local wisdom, collective creation of new knowledge, and the joint construction of effective tools in the same location where problems occur.

My argument holds that the Peasant-to-Peasant movements (but also other types of agroecology experiences derived from liberation theology) could well be situated in the second side of the equation. With clear goals in mind, they were able to jumpstart the ability of rural communities to use available resources, rekindle a network of human relationships, and restart solidarity, cooperation, and reciprocity that had been stifled by development practices and its Green Revolution technologies.

Through the exchange of knowledge, these movements have allowed peasants to revisit concrete solutions to concrete problems, freeing individual and social potential that were incapacitated by development (Robert & Rahnema, 2015). Peasant-to-peasant has revitalized relationship wealth, i.e., traditional wisdom grounded in the ecological particularities of the place. Thanks to the help of a horizontal exchange tool of agroecological knowledge based on the promoter, this wealth, made invisible by a discourse on poverty and lulled to complacency due to the mechanisms of modern aid, was mobilized, reactivating the potential to work together and preserve face-to-face personal relations. The imagination, the ability to create new knowledge, to experiment, and to encourage the wide distribution of wisdom, were factors that were reactivated by a methodology that is not new but rather recalls the community fabric that for eons has given people independence and security in the face of threats that could leave people in situations of precariousness.

In the end, the communal tradition has been reactivated by the social processes involved in agroecological expansion. In a world where what was free has gradually become merchandise, the commons were privatized, the monopolies were created, and the possibility of joint access to local wealth was banned, the previous cases teach us that it is possible to place the systems of production and consumption under social control, a process that includes granting access to seeds, technologies, tools, and wisdom. Distancing themselves from a system that stimulated individual profiteering, greed, and interpersonal competition while separating the individual from the community, the rekindling of community-based systems, encouraged by agroecological social movements, had the effect of promoting feelings of belonging to a social body, activating community linkages and mutual aid, and prompting peasants to think of themselves as "we."

The fruits of peasant-to-peasant can also be viewed by analyzing its architecture. This movement is about hundreds of thousands of families, even millions in the case of India, which tells us what social power can be when it becomes a mass phenomenon. It is difficult to quantify how many billions of dollars have been spent on research in order to create agronomic, zootechnic, and biotechnology packages and how many billions more in extension and in other misguided projects. Alternative experiences, on the other hand, are proof of the potential that resides in austerity, in becoming part of an organized body, the benefits of rehabilitating community spaces, and the advantages of relational structures based on mass participation and collective creativity. What I want to emphasize is the lack of efficiency—a pillar of development's economic rationality—of experts' monopoly on technology, as compared to the possibilities made available by open innovation of flexible techniques that can be imaginatively adapted to local conditions (Giraldo, 2016b). The peasant-to-peasant social movements are not only undermining outlandish development projects but they are also shining a light on how to build alternatives once the relational wealth of local communities is activated. Based on this networked architecture, communal wisdom can circulate, and new knowledge can be produced through hybridization, dialog, information recombination, and collective learning (Escobar, 2005).

This is not a minor matter. We should recall that the technologies of agricultural capitalism work in such a way that users' access to their design and manufacture are blocked (Harvey, 2003). The research and development departments of corporations or universities—often with public funds—undertake research and then transfer the technology to recipients who passively accept it. Peasant-to-peasant subverts this logic by putting experimentation and knowledge dissemination under the control of rural grassroots sectors, thus ensuring that the result is based on a dialog of wisdoms. Because of these types of network designs, it is possible, as Illich (2006b) believed, to strengthen social creativity and strengthen the linkages wrought by coexistence. This occurs once horizontally distributed knowledge is capable of bringing these tools into the service of persons who are organized collectively and not into the service of a corps of specialists.

I want to make three comments regarding this flow of wisdom that circulates through decentralized networks. The first idea comes from the Deleuzian thinker De Landa (2011) and his book *A thousand years of nonlinear history.* De Landa mentions that dynamism cannot fully emerge when a society's hierarchical components predominate over its horizontal components. Until the sixteenth century, he argues, the Orient was better positioned than Europe to dominate the rest of the second millennium. Yet the "excess centralization of decision making" and the dependency that this created "on the elites' skills," (Ibíd., p. 63) led to the European powers overtaking China and Islam on the world stage. "An inept sultan could paralyze the Ottoman Empire as no pope or emperor in the Holy Roman Empire could in Europe as a whole" (Kennedy, 1987, quoted by De Landa, 2011, p. 63). What I am trying to get at, following de Landa, is that excess centralization and hierarchy and the rigid vertical communication flows—i.e., that found in producing technological knowledge in market capitalism or to a greater extent in state capitalism—are much less efficient than structures in which people are not just passive information recipients. Greater efficiency is produced when all recipients simultaneously become emitters and creators of local knowledge (Escobar & Osterweil, 2009).

Traffic apps for smartphones, such as Waze, or platforms such as Wikipedia have shown—from a different perspective—the enormous advantages that accrue from unfettered tools of the "commons" that allow everyone to improve them. It is very difficult for hierarchical designs, where just a few people have control over knowledge, to compete with the social power of open and participatory designs, whose networks prize creativity and information sharing. Similarly, peasant-to-peasant has shown the immense possibilities of a grassroots network in which members organize to form a web pattern that connects them without imposing any uniformity. These are expanding networks that grow exponentially due to the encouragement given to experimentation and decentralization. In the final analysis, they contradict neoliberal dogma regarding the efficiency of the free market, in which operations are based on top-to-bottom, technology-creation systems. The polycentric network and internetwork designs of post-development demonstrate the importance of widely disseminating knowledge and the potential for economies of solidarity in terms of the exchange, blending, and reuse of local wisdom (Escobar & Osterweil, 2009).

My second comment involves autonomy, a principle of post-development practices. As opposed to the disabling dependencies that development planners tend to generate, an alternative path would need to increase people's control over their lives. Yet, the experiences described herein show that it is not a matter of appealing to absolute autonomy or radical anarchy. As Jean Robert (2012, p. 177) says, this is about seeking "the conditions for achieving autonomy complemented by heteronomous resources," under the principle of "ever more autonomy." This has occurred in the Peasant-to-Peasant movements. Rather than seek isolation, they have created positive synergy between autonomy and heteronomy. The histories discussed herein demonstrate that it is possible for some participants to play a catalyst role by facilitating connections among members that later become mutually stimulating. There has been a satisfactory mixture of exogenous components with organizational components of grassroots producers. This has allowed atomized efforts to come together, stimulate interactions, and create positive feedback, amplifying agroecology on a territorial scale.

The third component refers to the importance of unpredictability. Although these processes have had simple and decentralized beginnings based on hierarchy—such as Palekar the advisor of the Zero Budget Spiritual Farming—or active planning—such as the Cuban case—we see that complex entities have later developed in which multiple actors interact dynamically and follow local rules instead of top-to-bottom decision-makers (Escobar, 2005). When this occurs, the movement may grow in any of a number of unforeseen directions. We know where the starting point is, but not the finish line. It would have been difficult to imagine that a meeting among a few rural producers from Mexico and Guatemala would later have such a decisive impact on Cuba, Mozambique, or Malawi or that the Karnataka movement in India would expand to so many states and even to other Asian countries such as Sri Lanka and Nepal. The mathematics of complexity has taught us about the butterfly effect, i.e., how small disturbances can produce spectacular changes at other levels. Nonetheless, for this to occur using the peasant-to-peasant mechanism, there must be a good combination between both the hierarchical and network components. It is true that there is a technical team in Cuba and Nicaragua that each carries out well-planned organization. Nevertheless, once volunteer promoters take over, the expansion develops more spontaneously. Horizontal self-organization is created and gathers strength up to the point where the network-like components prevail over the central control and verticality that existed at the beginning of the process. The case of India best illustrates how growth happens chaotically and in a barely controllable way (Giraldo, 2016b).

I believe that the social processes within agroecology teach us how to imagine an economy in post-development that can be integrated with new social relationships and controlled by society, and not the other way around as in the case of economic liberalism. In contrast to development's harmful practices, which can only portray the small producers and indigenous people in terms of a lack of profitability, efficiency, productivity, capital, and technology, i.e., the elements that the professional planner has in abundance, the post-development of peasant-to-peasant shows that it is possible to count on the varied wealth of rural residents, such as the strength of relationships based on values of solidarity, the power of community life, and the ability to innovate, create, and jointly produce new knowledge. This is eloquent living proof that instead of being included in a market economy based on economic rationality, people must act grounded in the local, through self-organization, self-management, self-supply, reciprocity, social redistribution of economic gains, exchanges with other people in the vicinity, and rekindling the numerous alternatives that individuals have to live a shared, simple, and diverse life (Robert & Rahnema, 2015).

Further, as Leff (2014) might say, the Peasant-to-Peasant movements are examples of a decolonization of knowledge that reinvents new territorialities while territorializing specific practices and ways of inhabiting. These mobilizations also seek to achieve the re-peasantization of the countryside while reviving ancient identities, new collective identities emerge. Thus, they cannot be solely viewed as a way of resisting dispossession and deterritorialization by globalized agro-capitalism. They are also true paths for defining ways of existing and re-existing (Porto-Gonçalves, 2002). They exemplify the disputes over the meaning of agriculture, stripped of economic rationality and domineering modernity, leading to what we currently call agroecology, a total resignification of rural people's lives, far from all forms of exploitation, to confront the disaster created by years of technology transfer that has brought ecocide and culturecide.[9]

Agroecology and Cracking Away at Capital

One of the most heated and interesting debates within the left in recent years has to do with the role of the state in social change. John Holloway (2005) reminds us that throughout the twentieth century, the debate centered on two viewpoints: undertaking revolutionary guerrilla warfare to conquer state power and accessing state power through democratic elections. Whatever the choice—or perhaps a combination of both—conquering state power was never questioned by either side as a way of transforming society. In Latin America, the first option wrote its epitaph after the signing of the peace accords between the FARC and the Colombian government. The second option lost some of its appeal given the disappointment caused by a boom of progressive governments during the first decade of the twenty-first century.[10] Many people who had pinned their hopes in these governments began to doubt

[9] Much research awaits in order to address questions such as the tensions that arise when agroecological experiences have commercial ties with the capitalist market. To what extent and under what conditions can heteronomy and autonomy generate synergistic relations that help strengthen these processes? What is the role of agroecological expansion in weakening patriarchy? At what point does agroecological growth begin to have an effect opposite to the one sought?

[10] The phenomenon began with Hugo Chávez's rise to power in 1998, followed by the election of leftist governments in Brazil (2003), Argentina (2003), Uruguay (2004), Bolivia (2005), Ecuador (2006), Nicaragua (2006), Paraguay (2008), and El Salvador (2009).

the role of the state in transforming their countries. What they saw had been foreshadowed by the Zapatista movement in Mexico: The state is part of the global capitalist structure and "any government that carries out any significant activity against the interests of capital will be faced with an economic crisis and capital flight from state coffers" (Holloway, 2005, p. 17).

What various social movements saw during the "hangover" of years of Latin American progressive policies is the capitalist nature of the state from whose grip is it impossible to withdraw. No matter what the political inclinations of any particular government may be, the state finds it impossible to cease encouraging the reproduction of capital.[11] This is not to say that it is impossible to bring changes. Living conditions for very many people can be improved and the effects of capitalism alleviated, which for vast grassroots sectors represent a fundamental change in daily life. What we need to realize is that progressive governments are not the same as neoliberal ones and thus cannot be treated in the same way. Yet the lesson we need to learn in all its brute radicalism is that we cannot expect a change in the capitalist system within state institutions and, thus, our dreams cannot be dropped into the ballot box.

A slow change within the left has occurred in terms of need to crack capital. As we have seen, the capitalist system is a vast wall that seems impenetrable and undefeatable. It seems impossible to knock down, and an increasing number of people have lost faith in bringing it down from within the state. A much more practical but less romantic option is to crack the wall, weakening its foundations from below, and, within the same society that we reject, create fissures that weaken the base that sustains the system (Holloway, 2011). Capitalism is nothing more than a certain social relationship, and, therefore, it is impossible to subvert it in any way other than changing social relationships, directly changing the fabric of social life. This is what I believe that the Peasant-to-Peasant movements and other processes that disseminate agroecology are doing. Not waiting for changes to come from the agriculture ministries or from rural development agencies. Rather, they are distancing themselves from state power structures and have chosen to build a different type of relationship from below, based on different principles. We are talking about the emergence of an autonomous thinking within social movements, where there is growing disbelief in public policies as a way to set order to and transform a society.[12]

Naturally, some still believe that eventually a "really good" government will be elected and change conditions leading to a new, more just world. It is difficult to cease believing in the power of institutions because it implies, as Esteva (2017) has noted, doing away with the wish to be governed through representatives and hierarchies of control. In spite of the growing evidence that a true democracy is

[11] James O'Connor (2001) states that it is hardly an exaggeration to say that all internal state functions aim to provide suitable conditions for the production of capital, i.e., making sure that the collective interests of capitalists are satisfied.

[12] See Vergara-Camus and Kay (2017) for an excellent review of the disappointment caused by progressive Latin American governments in agricultural matters.

unworkable in a centralized nation-state, the idea of subordination and the delegation of power to others continues to persist. For example, some agroecological movements believe that agroecology needs to be supported by national and multilateral institutions whose focus is on sustainable development. Nothing could be more undesirable. Agroecology is constrained by the logic behind development projects, having shown that it can only grow when it acts "without asking for anyone's permission" and when promoted by grassroots organizations themselves. There is much evidence. One of the stellar examples of self-organization without state support has been the Zero Budget Spiritual Agriculture, inspired by the Gandhian maxim, "if you want to change the world, change yourself," and by the Zapatistas in Chiapas who exercise direct democracy and self-determination based on the principle of "lead by obeying."

Other people, less convinced of the path of autonomy, yet fed up with political parties and governments, have implemented actions by themselves, without waiting for permission from higher levels of authority. Disgruntled with petitioning the state as well, they have chosen a different path supported by their organizations. With less discursive clarity and at times moved more by a survival instinct, they are exercising their power without ceding it to the state, as they practice self-government, i.e., leading their own lives within a communal consensus (Esteva, 2011). There are still other experiences, perhaps the most numerous, that carry out pragmatic autonomy. Without completely rejecting the state, they use it for their benefit by taking advantage of its programs. In this sense, communities are not simply victims of development nor passive actors. There is also an art of resistance, which means that communities often accept subsidies and projects but adapt them to their own ends. They thus simulate that they are still controlled by the state when in reality the objective is to keep government intervention at bay.

Whichever path is chosen, as organizations implement their practices, the option of being "in the political realm" rejects political parties and centralized planning. Heterarchical trials in exercising power are emerging that chip away, even slightly, at institutional biopolitics. Agroecology or, more precisely, its social processes have been opening fissures in the monolith of capitalism through practices that displace capitalist social relationship and in their stead build relationships that recover the power of local wisdom and community values. We hardly ever see a yearning for the dictatorship of the proletariat, single-party politics, state bureaucracy, centralism, industrialization as a central tenet of communism, and state ownership of the means of production. Although some vestiges refuse to die, in countries around the world *politics of place* are arising that look askance at the grandiose policies of the Revolution (Escobar, 2005). People everywhere are choosing the preferential option of self-management, autonomy, and joint ownership of the ways and means of production. These movements refuse to see their dreams reduced to replacing market capitalism with state capitalism. Quite the opposite. They are anti-capitalists insofar as they know that the only way to change the world is by implementing another type of social relationships.

We are not on the cusp of a new style of anarchism because these movements do not seek or desire to abolish the state. They are channeling their efforts in other directions. In the spirit of the Zapatistas, their objective is not to "change the world" but rather to "build a new one." Their goal is not reformist because they do not hope to change institutions from within. Rather, at the local level, they are experimenting with new ways of living, which may be much less pretentious than the great Revolution, but much more realistic, because they are in practice creating a grass-roots, utopist practice of concrete action. When small producers cease buying agro-technological inputs from the multinational companies and begin experimenting on their own plots with previously forgotten wisdom, creating new knowledge while dialoging with other small producers who have traveled the same path, they are indeed creating a new world without waiting for state institutions. With this perspective, even within a capitalist society, very different relations can be created where there is no acceptance of exploitation of humans by humans or society's exploitation of nature. This is a construct where inhabitants act in a voluntary and cooperative manner to recover control over their lives. In this self-organizing setting of noncommercial exchanges, state institutions are of little use. Thus, rather than wanting to destroy them, movements either use state institutions in an anti-hegemonic way in their interactions or completely ignore them.

In recent years, a lively discussion within other-world movements has centered on the belief that a postcapitalist world begins by opening cracks in the system, which implies stepping away from ongoing development schemes and heading down a truly new path. Armando Bartra's (2008, p. 161) image of this transformation is the following: "a durable world-system whose vitality has been sapped and is in decline is simultaneously being premeditatedly dismantled in different ways and in many places, with the goal of gradually replacing it with something else." The question being asked by those involved in new forms of struggle is not how to occupy the state and replace its leadership, but rather how to deepen, make visible, and connect the fissures that are emerging in different places and moments, in order to progressively fracture the capitalist structure (Holloway, 2011). In this sense, agroecological experiences, the mayor ones discussed herein, other smaller ones supported by liberation theology, and still others aided by civil society organizations, which are widely scattered and doing their work silently and discretely, should be recognized as being precisely part of those fissures. They are simultaneously unweaving the fabric of agro-capitalism and finding alternatives to those based on technical progress and capital accumulation.

We should, however, sound some warnings. There are quite a few defenders of agroecology and the new agrarian movements, including many academics and members of NGOs, who continue to be closely associated with development projects and who still yearn to be governed. Due to naiveté or excessive pragmatism, they believe it is possible to institutionalize agroecology and disseminate it with bureaucratic support. They fail to foresee that with the recent inclusion of agroecology in the agenda of international organizations such as the FAO, post-development achievements might be corrupted. Undoubtedly, governments will try to create new bureaucracies to provide agroecological services through the state, opportunistic NGOs, transnational companies, and projects funded by international foundations and organizations. Similarly to what occurred during the development era, a possible institutionalization of agroecology will create clients of projects, after which time we run the danger of seeing rural inhabitants become objects of professional agroecological services provided by experts (Giraldo & Rosset, 2017).

A second warning: We must not overlook the imprints of power in modernity nor the economic rationality lurking within agroecology. Both the dominated and the domineering share an economistic symptom that is expressed when people in resistance continue to measure agroecology in terms of productivity, output, and efficiency, in order to compare it with conventional agriculture. Although it is true that in some circles it is only possible to speak the tribal language of modernists and that scientific research about these topics is very useful in political polemics, it is also true that agroecology continues being conceived with the same terminology of capitalist modernity. This includes anthropocentric concepts linked to sustainable development discourse, including agroecosystems management, natural resources, ecosystemic services, or environmental management. We should recognize that a problem with these resistance movements is that they are inevitably arising within the system that oppresses them. As Santiago Castro-Gómez (2015, p. 18) has stated in his dialog with philosopher Žižek, "the resistance will have to be coordinated "in" and "from" the techniques and tactics developed by power itself." Indeed, "The subjects in resistance really play with the same rules that they struggle against, given that they are the product of disciplinary norms and bio-political technologies that go hand-in-glove with the rise of capitalism." In our case, the yearning to be governed, economicism, and anthropocentrism are still latent in agroecology and arise as a symptom when many activists and scientists try to resist while using the same terms of reference as those that built globalized agro-capitalism.

Even so, it is difficult to be judgmental. As we have discussed at length in previous chapters, so many years have gone into creating truths about agriculture, human habitation, and nature that it is reasonable for us to feel strange as we cease identifying ourselves with a "Master-Signifier," as Lacan would say, and take our destiny into our own hands in a direct exercise of power. Nor is it easy to escape the imaginary of developmentalism. Modern beliefs involving the separation between culture and nature, subject and object, individual and community, cause and effect, technological fetishization, the world perceived as an orderly place, and the certainties of progress bewilder the resistances and continue imprisoning us in thought processes based on the dualisms of modernity and the desire to be servants. As Lacanian psychoanalysis teaches us, no cure will work if it overlooks our ontological illness or fails to overcome the symbolic order that structures our social relations. We ought, then, to recognize that utopias must acknowledge the ideological structure in which they take shape and accepting this is the first step to build diverse worlds that are distant from the regimes of truth that currently sustain our actions.

I believe that agroecology has taken several steps at cracking away at capital and is showing nonagricultural movements how to move toward post-development. The key factor is strengthening knowledge based on the place and wisdom to flow throughout horizontal internetworks or though face-to-face contact, without commercial transactions or state intervention. The objective is not to disappear the state as it is, in Gustavo Estevas's (2008, p. 9) words, "withdraw political power from throughout the state apparatus and assign it solely administrative coordination and service functions," as well as "salvage the principle of democracy: government by the people, power of the people," something very different than state power (Esteva, 2011, p. 137). I believe another key factor must be handled separately and carefully: the ability to include the economy, technology, and community relationships within the cycles of ecosystems.

Chapter 6
Reinhabiting the Earth's Crust Through Agroecology

> Intelligence took us from being a simple, defenseless animal to a geological transformer. And wisdom will guide us toward integration with the cyclical processes that make life possible
>
> Giovannie Soto-Torres

In the first chapter, we saw at length how the metaphysical thought that informs agroextractive technology has been formed by ignoring the conditions that make possible the reproduction of life and people's cultural diversity. Metaphysical thought built a world removed from biological materiality and peoples' cultural diversity. Then we discussed why agribusiness technology, based on economic rationality, was not designed to inhabit an overabundant earth or respond to human needs. Rather, it was created and re-created to further capital accumulation. With these two aspects, plus others that we have seen along the way, we have tried to explain that agroextractivism is yet a symptom of a civilizational crisis. This crisis has its origins in the fact that a technological platform was built that disregards nature, subordinated to promoting the interests of big capital. For this reason, if we are interested in building other agricultures, not for developing ourselves, or progressing in any particular direction, but for inhabiting the world that we have disinhabited, then we need to imagine other technical principles in tandem with the ecological and cultural conditions of specific locations, so that a dialog between technical expertise and nature can take place.

As we did in the previous section, here we undertake a dialectical exercise with the predatory technology of extractive agribusiness, by rejecting its logic and starting anew with an opposite perspective. Instead of focusing on economic profitability and how to adapt ecosystems to productivist thinking, in post-development we

---

A version of this text was published as Giraldo, O. F. "Agroecología y complejidad. Acoplamiento de la técnica a la organización ecosistémica," *Polis Revista Latinoamericana,* 14 (44): 277301, 2015.

O. F. Giraldo, *Political Ecology of Agriculture*,
https://doi.org/10.1007/978-3-030-11824-2_6

begin by understanding the specificities of ecosystems and then imagine how we insert technology into the rhythms of life. The agriculture that has prevailed for millennia teaches us that we first need to concentrate on the ecological possibilities and potential of the place we inhabit and then focus on how to integrate technique into nature so that the transformation of ecosystems will not reject—or hinder—the continuity of the web of life.

We shall proceed in this section in the following order. We will focus on physical aspects by reviewing some of the concepts of the systemic paradigm of biological sciences. We will cover the epistemological progress made regarding the ontological question of life. We will begin our discussion based on a theoretical foundation that covers the conditions for life to be possible on this planet and then consider the possibilities of integration, harmonization, coordination, and fitting in with the vital substratum that we inhabit.

We begin by studying an interpretation of the theory of evolution and then cover the epistemic foundations of a technical approach that is compatible with the biophysical materiality of Earth, which is, in the end, one of the greatest challenges in building post-development as we approach a world beyond capitalism.

Agriculture, Coevolution, and Natural Drift

Francisco Varela and Humberto Maturana have made intriguing amendments to evolutionary theory that, as we shall see, lay an epistemic foundation for agroecology, mainly for exploring what other technical possibilities exist for habitability. Varela and Maturana have called their approach "natural drift," which has many aspects that go beyond our discussion here, and so we concentrate exclusively on those that contribute to the discussion of agroecological techniques. To understand evolution by natural drift, we begin by examining the critique made by these biologists of neo-Darwinian evolution. Their main critique is of *adaptation*, a concept that posits that survival is determined by an organism's ability to adapt to its surroundings. According to this neo-Darwinian notion, natural selection works by choosing those adaptations that tackle the environment most effectively. The fins of fish, for example, are what they are because they are best adapted to an aquatic environment. The problem with this description, they say, is the way of perceiving the medium, as if it were an independent and pre-given space that imposes *restrictions* to which species must adapt if they are to survive (Varela, Thompson, & Rosch, 1997).

Varela propose a refinement based on two substantial changes. The first involves shifting from a *prescriptive* logic to a *proscriptive* logic, i.e., from the idea that "whatever is not permitted is prohibited" to the idea "whatever is not prohibited is permitted" (1997, p. 227). The second change is to cease viewing evolution as an optimizing process and rather more as *satisfactory*. Returning to our previous example, fish fins, then, are not optimally adapted to water but rather are a viable morphological aspect among many others that were possible during the course of evolutionary history. Fins are not optimal for swimming but rather "satisfactory," insofar as their structure is an adaptation to the aquatic environment. Yet, many other structures might have satisfactorily fulfilled the conditions necessary for the same aquatic environment.

For natural drift thinking, evolution depends on the existence of "a structural connection" that allows some of the multiple viable trajectories to be "satisfactory." This is a shared history between organism and environment, in which the necessary structural coherence has occurred. Maturana and Varela (2003, p. 82) state, "Evolution occurs because the organism and the environment persist in a continuous structural coupling. At each moment, all organisms, ourselves included, work as we do and are where we are due to a structural coupling." By incorporating phenomenology in biological science, they conclude that the environment is not "pre-given" nor imposed on organisms that then have to adapt to preexisting conditions. Rather, both the environment and the organism evolved together (Varela, 2000).

A good example of coevolution is the structural coupling between bees and flowers. Bees can see something that humans cannot, i.e., flowers' ultraviolet reflectance. Through ultraviolet reflectance, flowers attract pollinators and thus guarantee their reproduction, just as bees distinguish flowers at a distance in order to gather food. These characteristics are explained as a history of connection, "in which the traits of plants and the sensorial/motor characteristics of bees evolved jointly" (Varela et al., 1997, p. 234). Natural drift posits that this connection is "responsible" for bees' ability to see ultraviolet light as well as for flowers' ultraviolet reflectance. In this case as in all evolution, a mutual specification exists between the organism and the environment. Yet this does not occur because of the optimal adaptation of a species to a preexisting world, but rather because of a *satisfactory* process of codetermination in which the bee and the flower evolved in a reciprocal manner.

Maturana and Varela emphasize that what is in an organism's surroundings, in this case the world of ultraviolet flowers for bees and the world of pollinators for flowers, exists due to a history of structural coupling. Yet this coupling is far from being perfect. A satisfactory and viable coupling merely depends on an interaction that facilitates the ongoing integrity of the system. This is a proscriptive logic since all actions are permitted as long as they do not violate the system's only prohibition, i.e., the maintenance of the system's integrity must not be violated. If the structural connection is interrupted, the restriction is violated, and the system as a whole will be destroyed.

This overview of the theory of natural drift has profound consequences for understanding the relationship between nature and agroextractivism and the agroecological conditions that are necessary for habitability in the context of post-development. Maturana and Varela state that coevolution exists because of a common history between an organism and the environment, where the connection between both has brought stability to the relationship. In the case of human beings, any explanation of our permanence in the world cannot skirt the issue of how other biological organisms have survived, and so we need to delve into the intrinsic characteristics of their evolution.

Anthropologist Clifford Geertz (1991) states that the radical difference between human beings and other animal species is that our evolution occurred because of culture. We are the species that most depends on extra-biological mechanisms to orient ourselves in the world. This is because our genes, as a source of information, are weak determinants of human behavior. A bird, for example, can intuitively make its nest; there is no genetic information available to guide humans in building a house. Unlike any other animal, we need non-biological sources of information. Nevertheless, Geertz says, culture was never added to a completely finished animal. In other words, a biological evolutionary process did not exist first, and then culture appeared afterward to round out that animal. We need to understand culture in a different way, as an integral part of what makes us human. For Geertz, human nature depends on culture. We became human beings due to the cultural structure and the meaning systems that we created. Without culture, he insists, we literally would not exist in this world.

Continuing with natural drift theory and the role of culture as a differential element in human coevolution, there is a necessary structural coherence between a living being and environment, such that culture, as an inherent characteristic of an animal's evolution, had to be created in compatibility with the environment. Had it not been done in this manner, survival of the species would not have been possible. Humberto Maturana (2007, p. 69) states quite emphatically, "We are as we are in coherence with our environment and… our environment is as it is in coherence with us; when this coherence is lost, we do not exist."

If we accept a radical interpretation of Maturana's thesis, culture, seen from the long lapse of time implicit in the species' permanence on Earth, must have been formed, created, and re-created in permanent coupling with nature. Obviously, this coupling did not successfully occur in all cultures. We need only recall the tragic history of a few civilizations, such as the Babylonian, which was unable to overcome the effects of growing soil salinity, or the Roman Empire, which could not avoid soil erosion (Ángel, 1996). The fact that the human species persists can only be explained as the result of some measure of coherence that has existed between culture and nature, which over the past 10,000 years has included cultivation of the soil.

Although agriculture is very recent in the evolutionary history of our species, it is an inseparable part of the coevolution of our culture and environment. From the Neolithic period, the agricultural revolution created the sedentary nature of human life on Earth, which implied an entire biocultural transformation. People learned to coexist with nature in a radically different way than the nomadic gatherer that preceded us. The nomads' constant movement slowly gave way to inhabiting an appropriated and rooted place that surrounded crops and domestic animals (Giraldo, 2013).

Consequently, a coevolutionary process emerges that is the result of humans' intervention in nature. Some seeds and a handful of animals were domesticated in a historical process of coupling between culture and ecosystems. After domestication of food plants such as wheat, barley, quinoa, corn, potatoes, rice, and millet, human intervention became necessary for their continued existence. In a manner similar to bees and flowers, the ten-millennia history of agriculture should be understood as a coevolutionary, ecocultural process with reciprocal interaction between culture and nature. This was not the adaptation of culture to a preexisting environment. Rather, it is a more-or-less satisfactory history of codetermination in which cultures, animals, and domesticated crops coevolved.

During this process, native populations, on the one hand, inhabited a nature transformed by agriculture and developed a profound understanding of ecosystems and a treasure of wisdom about how to grow crops and graze animals according to the specific contexts of the place chosen to inhabit. On the other hand, native populations contributed to the planet's biodiversity through the domestication of 5000 crops, 1.9 million varieties of vegetables, and 40 livestock species (ETC, 2009). They emerged from an ecocultural coevolution that dates from the origins of agriculture and drastically changes the coevolution of human beings and their environments during the prior 200,000 years of history. During this period, nature coevolved together with technical know-how. An example of this is the immense treasure of biocultural diversity that has co-emerged during the last ten millennia.

The culture and environment inhabited by human beings are not determined separately and even less so after agriculture was created. Their history is common to both. This view rejects the idea that the environment has been imposed on the human animal from without, such that humans have needed to adapt their culture to preexisting surroundings. Varela and Maturana's thesis implies accepting that nature, transformed by agriculture, is largely a creation of human beings, similarly to how human beings are the creation of modified surroundings. Both have evolved through a reciprocal process of mutual co-creation.

Since Neolithic times, nature inhabited by sedentary communities becomes inscribed, transformed, and affixed to people's cultural histories. The joint history that extends for centuries has a relatively satisfactory structural coupling, because the technique used to transform ecosystems facilitates the continuity and integrity of the agroecosystem. In other words, over millennia life was structured by a compatible symbiosis between agricultural intervention and ecological cycles.

This is a far-from-perfect process. We should not deny the environmental impact involved in implementing agriculture given the ensuing increase in human population. Ecological science tells us that natural population limits exist in order to maintain biotic stability within the ecosystem. But agriculture allowed human beings to transgress ecological laws by increasing their population at the expense of nature's limits. Due to agriculture, human beings were expelled from ecosystemic paradise, i.e., we no longer occupy an ecological niche (Ángel, 1996), and from that point forward, we refused to be constrained by ecosystems' self-organizing relationships.

Yet in an eminently technical matter, the coupling of agriculture to the environment occurred in a relatively satisfactory way to the extent that the survival of our species was assured over the years. If the modification brought by agriculture had been based on a predation of soils and the destruction of biodiversity, the necessary coupling between species and environment would not have been achieved, and the circular interaction and natural regeneration of transformed ecosystems would have been interrupted. Biologically, this would have led to the extinction of human beings on Earth (Giraldo, 2013). For this reason, the coevolutionary process has been in some sense satisfactory, given that many feasible trajectories were possible, which are expressed in humans by their multiplicity of cultures. Throughout the centuries, these multiple cultures have allowed co-habitation between the human species and other multiple forms of life.[1]

This interpretation ceases to be accurate with the globalization of capitalism and its intrinsic agroextractivist technology, which acquired its full predatory impetus at the end of the nineteenth century, as we saw in Chap. 2, when a historical freefall is consolidated, first in Europe, when it begins to shun coevolution during the Industrial Revolution. This blunder against nature committed by Western civilization can be explained by the analogy of the cancer cell and its miscommunication with its extracellular surroundings.

Recalling natural drift theory, a biological system's sole restriction is that the action of its component parts cannot violate its integrity, because if the structural coupling is disrupted, systemic continuity is lost. A cell is considered cancerous when its communication with the organism is suddenly interrupted. From one moment to the next, it violates its only restriction, i.e., that its existence allow for the survival of the system. The cancer cell is oblivious to this sole law because it has ceased to communicate with its surroundings (Varela, 2004). This analogy is useful because the relationship involved is similar to what agroextractivism does to the earth: It strips the substratum of what it needs to exist. Just as the cancer cell is incapable of establishing communication with its surroundings, agroextractivism is incapable of communicating with life forces and adhering to the only prohibition imposed on it.

Over hundreds of generations, the agriculture that today we call agroecological—to differentiate it from its industrial counterpart—built its techniques, guided by the ecological characteristics of inhabited places, in a permanent dialog with the imminences of life. In contrast, the agroextractivism that arose over a few decades abruptly broke off communication with its surroundings, making it impossible for the whole system to survive. This essential difference, seen in the light of the coevolution of species, can help us understand the way agricultural practices must be attuned in order for human beings to continue on this Earth.

The natural drift perspective has provided lessons for agroecological techniques that can be summarized as follows. First, many diverse ways, open to the imagination, exist for creating the necessary conditions for habitability through agroecology. In fact, Earth's original peoples invented and reinvented many of the techniques used in coevolutionary, ecocultural processes. In many of these cases, work is not about inventing anything new but rather rediscovering the connection between cul-

---

[1] Humans' short transit on this planet has caused a massive extinction of biodiversity that began in the late Pleistocene and accelerated during the Neolithic. Nonetheless, the impact of our species on the ecosystems of these periods is insignificant compared to the environmental disaster of our modern era (Broswimmer, 2005). This is why we should endeavor to understand how numerous groups of original peoples were able to connect with their environment, not in a harmonic and balanced way, but with tension and conflict, with wise choices and errors.

ture and environment that today has become muddled by the teleologies of progress and development. Second, the aphorism "what's not prohibited is permitted" means that human creativity can soar as long as it respects nature's only inexorable restriction, i.e., no action may interfere with the integrity of the substratum that is necessary for the agroecosystem to survive. Paraphrasing Maturana and Dávila (2013), we can do anything that our imagination deems as long as we respect the structural coherencies of the place where we are imagining. This means that the path of human creativity for intervention and technical innovation is open, as long as we do not hinder the reproduction of life's complex interrelationships that inhabit us and that we inhabit. We must learn to listen to the physics of the biosphere so that all techniques coupling to the interrelated life system that characterizes our beautiful, living planet.

To improve our understanding of this vital technical principle, we need to explore in detail the conditions that make life possible on Earth. We next take this step to further our understanding of the epistemic principles of agroecology's technical activity and the transitions to post-development and post-extractivism.

What Is Life? Autopoiesis and Agroecology

The essential difference between our beautiful planet and the rest of the known universe has to do with the special conditions that make it possible for Earth to harbor the magic of life. This unique trait that we take for granted is not easily and precisely defined.

Varela (2000) uses an example to describe this difficulty. He says that in spite of knowing with certainty how to distinguish a living organism from one that is not alive, we would not be so confident in arguing why a tree, mosquito, mule, worm, coral, and human are alive, while a radio, computer, robot, or sea tides are not. Varela says that even though we might be tempted to answer that movement, growth, reproduction, reaction to stimuli, or energy transformation are hallmarks of all living organisms, someone might point out that trees do not move or show signs of growth in the short term, while sea tides do grow at regular intervals. Likewise, a mule or a child is incapable of reproducing; trees and corals do not flinch when pricked, while a radio or a robot can function through the transformation of external energy in ways similar to living beings.

There is, however, one shared trait that the tree, mosquito, mule, worm, coral, or human can do that the radio, computer, robot, or sea tides cannot, i.e., the ability to regenerate themselves from their own interior. A tree loses its leaves in winter but can re-create them the following spring, and hair that falls from an animal can grow out again. Varela (2000, p. 26) posits that the quality that distinguishes all living organisms is that their "tissue undergoes a continuous process of destruction and regeneration due to some type of activity in their interior." A robot, a computer, and a radio are unable to regenerate themselves from within, while living beings are able to renew their own components, using external exergy to maintain their structure.

This macroscopic explanation of what is alive is a comparison to what Maturana and Varela have called, on a unicellular level, "autopoiesis." This term, newly minted by biologists in the 1970s, helps to define the process of "self-creation" that cells constantly undertake to regenerate themselves. Through "autopoiesis," Maturana and Varela emphasize the autonomy of living systems as the specific quality that differentiates a living unit from a nonliving one. In their own words, "living beings are characterized because, literally, they continuously produce themselves" (Maturana & Varela, 2003, p. 25). In other words, what distinguishes a living unit from a nonliving unit is its ability to self-create permanently, i.e., they "produce themselves" by means of the components that their own structures generate.

Neurobiologists Maturana and Varela believe autonomy is the factor that best defines life. Our beautiful blue planet that floats in the cosmic immensity can be identified by a very peculiar characteristic, i.e., its living networks have the extraordinary ability to continually produce themselves. In living chains, each product carries out a function by participating in the production or transformation of other components of the network, such that the entire structure "makes itself" autonomously. Life works like a closed and self-organizing circle where, as Capra (1998, p. 181) says, "order and behavior are not imposed from outside but are established by the system itself."

The theory of "autopoiesis" is very precise in explaining the pattern of self-production that characterizes all living entities. This involves a "system of elements whose interrelationship produces nothing other than the entity itself" (Escobar, 2012, p. 7). For living things, the phenomenon is always the same: elements come together to form a unit of circular interaction. It involves an entire process of organization, or, better, self-organization, by which a system of closed, circular operation is created, in which the unit creates components whose only objective is to produce a network of relations of the unit that produces them (Capra, 1998).

At this point, we can point out some of the implications of any technique that aspires to ground itself in the conditions that make life possible on Earth. If we understand that life works in a circular system, techniques for inhabitation must abide by the self-organizing cycles of ecosystems. If nature speaks the language of cyclicality, techniques must not be linear in their interventions. If nature is cyclical, if the function of all components of a trophic chain rest on transforming the elements of the network, such that the system operates in a circular organizational process, a technological platform, such as agroextractivism, cannot function as a linear system, at least not if it wishes to maintain the life system on which it depends to survive.

As we have emphasized, our current environmental problems exist because capitalist civilization built a social order independent from nature. For this reason, linear output and consumption that do not return to their point of departure and do not reintegrate into the earth in a cycle of continuous return are truly incompatible with the reproduction of the planet's living relationships.

The consequences for the Earth of agroextractivism's linear process are very similar to what occurs with a biosystem structure when its life is shut down. "What is destroyed when a living system is dissected," says Fritjof Capra (1998, p. 99) "is its pattern. Its components remain where they were but the configuration of their interconnecting relationships, its pattern, has been destroyed and so the organism dies." What is important in a living system is the preservation of the interactions and relationships among its parts. If this pattern of ordered relationships is interrupted, the system loses the trait that distinguishes it from nonliving objects, i.e., its internal regenerative capacity. In simple terms, an organism dies when it loses its autonomy to renew itself.

A Fordist and irreversible intervention, such as extractive agribusiness, hinders the process of life, which, in terms of autopoiesis, implies the destruction of ecosystems' "self-creation." By imposing linear technology through intensive and highly mechanized monocropping whereby crops and livestock are separated and the soil's progressive fertilization is supplemented with enormous doses of chemical fertilizers, ecological stability is undermined through pesticides, and hybrid seeds cannot self-reproduce through free pollination (Bejarano, 2003). Thus, the ecosystemic cycle is hampered. This leads to a breakdown of the system's integrity aimed at continuing life. Ecosystems are autonomously organized in a nonlinear process; to reproduce itself, the system depends on continuing cyclicality. If this is hindered by interrupting the return of plundered elements to the earth, eventually this living network will die. For this reason, the fundamental principle says that all actions are permitted as long as the intrinsic cyclicality of nature is not interrupted and an organism's organization can maintain its autonomy.

Human creativity in pursuit of technological innovation must not be restrained. The idea is not to disengage from the transformation of ecosystems. Quite the opposite: We human beings have transformed ourselves by modifying our natural surroundings through technology. As Augusto Ángel Maya (1996, p. 51) says, "The solution to the environmental problem does not involve not transforming, but rather transforming well." In order to transform well, we need to heed the ecosystem's sole restriction, i.e., we must respect its cyclicity in order for life to continue.

The main element of agroecological technology, whose rationale has been understood for millennia because of people's trial-and-error approach, involves transforming ecosystems by coupling to nature's cycles. Thus, we are enjoined to understand that ecosystems do not produce waste because waste produced by one species is another species' food. The wastes generated by individual parts are continually recycled by the system in a holistic manner (Capra, 1998). If we examine this framework closely, we recognize that the techniques grounded in Earth's biophysical order must be integrated into the circularity of the planet's biotic recycling.

Altieri and Nicholls (2000) established six agroecological principles that are the fundamentals of a technique that is integrated into the cyclicity of ecosystems. The first is plant and animal diversification within the agroecosystem. Second is recycling nutrients and organic material. Third is managing organic material and stimulating soil biology in order to provide the best nutrients for crop growth. The fourth is minimizing water and nutrient loss by "maintaining soil cover, controlling erosion, and managing the microclimate" (2000, p. 29). The fifth is using preventative measures to control insects, pathogens, and weeds by promoting beneficial fauna, allelopathy, and a series of techniques developed by different peoples over millennia. The sixth is taking advantage of synergies and symbiosis through the interaction between plants and animals. The objective is to sustain a proscriptive frame of reference that favors highly specific creativity linked to "the place" but avoids interference with ecosystemic flows and cycles.

There are many examples of agroecological techniques, all grounded in the self-organizing cyclicity of agroecosystems without generating waste. Yet there are other principles related to the thermodynamics of life on Earth that need to be reviewed further.

The Second Law of Thermodynamics in Agroecology

We have stressed that life works like a closed system. This is true in terms of the self-organizing processes, but we need to include its interdependence with the surroundings in this analysis. The fact that a living entity is autonomous in its regeneration does not mean that it is independent from its environment. Indeed, the ecosystem is closed in its structure but open in terms of the energy and resource flows that are needed in order to maintain its structure. Thus, green plants use their roots to absorb water and mineral salts, which reach the leaves and then combine with the air's carbon dioxide to make sugars and other organic compounds. During this process known as photosynthesis, solar energy is transformed into organic energy, while oxygen is released into the atmosphere and can be used by animals and other plants to breathe (Capra, 1998). Green plants, whose biology makes photosynthesis possible, are eaten by animals that in turn are food for other animals, which when they die are decomposed by insects and bacteria in the soil that disintegrate them into nutrients. In a continuous recycling of organic material, green plants will then absorb these elements. During the process, what is waste for some becomes food for others so that the system recycles all elements and the ecosystem as a whole produces no waste. A single waste product, however, cannot be recycled throughout the food chain: thermal energy dissipated during respiration. Energy lost to the atmosphere is irreversible insofar as it cannot be recycled by the system. This means that matter circulates throughout a living system, but thermal energy is lost forever (Capra, 1998).

This physical phenomenon is known as "entropy" and is the basis of the second law of thermodynamics proposed by Rudolf Clausius in the mid-nineteenth century. According to this law, as a thermic phenomenon increases, entropy increases, but the resulting heat energy cannot be recovered.

Clausius formulated the law in terms of "waste" and "loss" because his research focused on improving productivity and minimizing the loss of energy in technology. However, this perspective changed at the end of the 1960s when Ilya Prigogine introduced a fundamental argument that energy dissipation was not a negative outcome as classical thermodynamics believed. On the contrary, he demonstrated how in living systems irreversibility was fundamental in terms of the impossibility of recovering energy (Capra, 1998).

Prigogine wondered what role the chaotic world described by thermodynamics plays in living beings, i.e., an unordered world that inevitably increases its entropic degradation (Capra, 1998). His answer was that the absence of equilibrium is an indispensable aspect of life. His conclusions, which earned him the 1977 Nobel Prize in Chemistry, state that in proximity to the state of equilibrium, everything is linear and there is no possibility of dynamism. In a state of non-equilibrium, there are many possible properties: matter is more flexible, new physical states appear, and there is a richness and diversity that are impossible to find in states of equilibrium (Prigogine & Stengers, 1984). For this reason, non-equilibrium and entropic irreversibility explain life. In fact, a living organism is never in a state of equilibrium. This state only occurs when its many metabolic processes are detained. As Capra says (Capra, 1998, p. 194), "an organism in equilibrium is a dead organism."

This paradox revealed by Prigogine means that non-equilibrium is a source of stability, since as we move away from equilibrium, the planet's richness and natural diversity increase. Thus, non-equilibrium, chaos, and the disorder produced by an unavoidable increase in entropy in the universe is a source of order (Prigogine & Stengers, 1984). Yet we should understand how this—apparently contradictory—biospheric ordering occurs. On the one hand, if the self-organization of ecosystems increases the entropic degradation of their surroundings through the dissipation of unrecoverable heat energy, on the other hand, organisms vacuum up "negative energy," using Schrödinger's term (2005),[2] or "negentropy" as Brillouin was to call it later, in order to maintain their living structure. This negentropy consists of low-entropy energy acquired from the environment, mainly sunlight, which green plants transform into biomass through photosynthesis. Biospheric stability is thus maintained between negentropic productivity and entropic degradation (Leff, 2004).

To better illustrate this phenomenon, we examine the thermodynamics of a tropical forest. The ecosystem generates entropy due to metabolic processes throughout food chains but also collects solar energy from the surroundings to transform it into biomass through photosynthesis. Thus, the biodiverse forest maintains its stability. The thermodynamic explanation of life holds that open ecological systems, such as the tropical forest, organize themselves though negentropic productivity. Through respiration, they dissipate heat energy and create increasingly greater entropy, which can only be compensated through the importation of negentropy and biomass production (Leff, 2004).

According to thermodynamics, a vital dialectic exists between entropy and negentropy consisting of a relationship of energy exchange that makes it possible for order to emerge from disorder. According to Prigogine, life involves a continuous flow of energy dissipation that combines with stability, such that the very constant environmental organization that we see at a macroscopic level is but an "island"

[2] Physicist Erwin Schrödinger (2005, p. 45) notes that a living organism avoids entropic degradation, i.e., the dangerous state of maximum entropy or death, by continuously withdrawing negative entropy from its environment. In his words, "An organism feeds upon is negative entropy. Or, to put it less paradoxically, the essential thing in metabolism is that the organism succeeds in freeing itself from all the entropy it cannot help producing while alive."

of order in an increasingly chaotic universe. Ecosystemic organizations are open systems that vacuum negentropy in search of order to the detriment of disorder in the context in which it interacts. In fact, the apparent contradiction is that, for a system to remain organized, it must increase the degree of disorganization in its surroundings (De Lisio, 2001). Yet there is no contradiction in this. In terms of energy, living structures can be explained at the intersection of entropic degradation and negentropy. This is so because, although heat cannot be recycled nor used as energy for life, at the same time, photosynthesis allows energy to be recirculated among different organisms by diverse metabolisms, minimizing entropy or emission of free energy, and by increasing the residency time of energy that is already within the system.

In other words, because challenging surroundings exist, live organisms organize themselves by drawing on free energy (negentropy) on which they are entirely dependent to live. The system itself extracts energy for its own functions, and due to the complexity of interactions and the great diversity of metabolic means overall, some interact with others to ensure that the indispensable energy needed for life is used by all (Schneider and Sagan, 2008).[3]

By means of multiple coevolutionary paths, life has organized itself by creating surprisingly efficient systems to obtain, store, and use energy (Toussaint and Schneider, 1998). Through adaptations, hybridization, degradation of organic material, cycling of nutrients, symbiosis, and competition, it has managed to delay the dissipation of useful energy, changing energy into biomass that is recirculated through food chains, thanks to transformations whose only waste is the thermic energy that is dissipated during respiration. Living organisms have structurally connected to the environment, maximizing absorption of negentropy originating in sunlight, while simultaneously minimizing internal entropy, in order to channel recirculated energy for its own preservation and reproduction. Life would not exist if it could afford the luxury of wasting energy. Every living organism must invest that energy in the best possible way to ensure its metabolic processes. For life overall, seen as a complex web of living beings that are fully interconnected and interrelated, its stability has been achieved, thanks to its immense ability to optimize the use of energy flows (Schneider and Sagan, 2008).

This thermodynamic explanation of life helps us to understand the effect that extractive agribusiness produces on nature. Oil-based agro-capitalism, linear in its output and distribution, degrades available useful energy by dissipating it as heat, i.e., low-quality energy that cannot be recovered. This is entropic maximization that can be explained by the fact that agribusiness contributes between 10 and 12% of global emissions that produce anthropogenic global warming (Lin et al., 2011). This does not include, of course, the ecosystemic simplification inherent in felling trees to make for vast expanses of barren land for agricultural and forest monocropping. The entire agricultural food system contributes half of all global greenhouse gas emissions (Grain, 2013).

[3]This section on thermodynamics in agroecosystems is based on a text by Andrea Vanegas (Giraldo & Vanegas, Unpublished).

When industrial agroextractivism transforms ecosystems into monocultures or vast expanses of flat land, the energy supply is cut off and the ecosystem's interrelationships (that existed before this intervention) are altered, reducing its ability to vacuum up energy. The effort to "discipline" ecosystems implies reversing the complexities of agroecosystems by reducing its energy flows. When an ecosystem is simplified, the same quantity of energy is not collected or stored by photosynthesis, and thus cycles are reduced, less material is circulated, interconnectivity is fractured, and there is much greater loss of nutrients and water (Schneider and Sagan, 2008). An agricultural system consisting of a single species is more vulnerable and unstable and can also weaken its ability to transform energy and thus reduce the possibility of reproducing itself autonomously.

Agro-biotechnology, with its metaphysical logic, has built linear processes close to equilibrium, ignoring the fact that nature is organized in states of non-equilibrium. Therefore, instead of creating order out of chaos, agribusiness exponentially increases entropy, thus contributing to the breakdown of the planet's ecological dynamic. By contrast, in agroecology, greater agrobiodiversity generates more energy and available biomass that are maintained within the system, in part due to the cycling of nutrients (Gliessman et al., 2007). This occurs because the increase in the diversity of species of plants, microorganisms, and animals means that there are a higher number of metabolic energy paths. Actually, when we mention agrobiodiversity as the guiding principle of agroecology, we are not referring to all the different species but rather to the number of metabolic paths available to degrade solar energy more efficiently (Kirwan, 2008). As an agroecosystem becomes more diverse, new paths are opened proportionally to transform energy flows, thus increasing complexity (Tyrtania, 2008). This diversity of biotic species and associated fauna makes the system more resistant, because redundancy is created for degrading energy (Mayer et al., 2014), and therefore it is less vulnerable to external disturbances. Thanks to its complexity, agroecological systems can better access solar energy, which is stored and used in a much more efficient way than is possible through monocropping and extractive agroindustry (Giampetro et al., 1992).

The living and dead organic material on the ground favored by agroecology becomes high-quality, stored, and available energy sources (Álvarez and Velásquez, 2013). Soil ecology liberates that energy for the growth of new plants and associated organisms (Addiscott, 2010) that increase the energy-residence time within the agroecosystem. Stated in other terms, agroecology favors the entire system's ability to postpone energy use through different metabolic cycles. The use of available energy is thus postponed, stored, and preserved to be rechanneled in the future, allowing the agroecosystem to continue as a dissipative structure over time.

Furthermore, the metabolism in agroecological systems is not limited to matters of cultivation. It also includes what Marx called social metabolism. This involves those energy-recirculation processes between croplands, domestic animals, and human beings, the latter of whom inhabit ecosystems that have been transformed with their own labor. Agroecology is broadly based on the use of biological energy coming from the work of human societies. Consequently, because of local consumption, an ongoing cyclicality exists between the use of biomass output derived from the transformation of available energy, the metabolism of human beings as constituent members of the system, and the recirculation of energy waste products, thanks to the many traditional practices of people who know how to return to the soil those elements withdrawn through their labor (Toledo, 2008).

Thermodynamic explanations help us understand that the workings of agroecosystems depend on continuous energy flows since, as dissipative structures, they work based on energy inputs to the system (Espinoza and Ortiz, 2014). The radical difference is that agroecology, by depending mainly on solar and human energy and reducing the use of nonrenewable energy inputs (González-Jacome et al., 2007), helps to compensate the irreversibility of natural entropy through the connection of human habitation to the energy necessary for the planet's ecological regulation. Further, the thermodynamics of life reveal how agroextractivism, being dependent on machinery, irrigation, chemically synthesized fertilizers, commercial pesticides, and an entire transportation network—and thus energy derived from hydrocarbons and other fossil fuels—(Dazhong and Pimentel, 1984), is disconnected from the biosphere's energy dynamics. This situation shatters the vital dialectic between the transformation of solar energy and the degradation of entropy in the universe.

Undoubtedly, industrial agriculture, having very low self-sufficiency and low negentropic productivity of biomass, dissipates high amounts of entropy in the form of heat. Agroecological systems, on the other hand, channel energy to the system's interior—by promoting a complex network of interactions and biological interdependencies, based on recycling of materials and waste—depend on local resources using human and animal energy, and use agricultural varieties adapted to the ecological conditions of specific territories (Gliessman et al., 2007), avoiding thermodynamic equilibrium. Although it is true that neither agroecology nor any technique based on the Earth's physics could reverse the unchangeable dissipation of entropy in the universe, agroecology involves a productive paradigm that harmoniously connects to biosphere thermodynamics and that in our turbulent times of environmental crisis—as La Vía Campesina suggests—contributes to cooling our planet.

In addition to the techniques developed by millenary peasant wisdom, more recent and diverse eco-techniques demonstrate how humans' immense creativity has been used to connect to ecosystemic niches. Thus, nature's language, autopoietic cyclicity, and negentropic productivity meet the symbolic order, which includes human ingenuity in modifying ecosystems in order to inhabit them. This involves all those techniques that are fed by low-entropy energy, particularly sunlight and the production of biomass from photosynthesis, which are incorporated into the cyclicity of diverse and self-organized ecosystems (Leff, 2004).

Agroecology views negentropy as an irreplaceable condition for the planet's ecological regulation and uses it as a part of the cycles of life. Thus, the various possible paths for human beings to connect structurally to available ecosystems—exemplified by Earth's cultures—are satisfied. The fundamental principle of all environmental organization, both ecosystemic and cultural, has nothing to do with monotony or uniformity but is based on diversification, plurality, multiplicity, and difference.

This is the reason why, thanks to biocultural diversity, multiple communities throughout the world have been able to connect with the environment during their coevolutionary processes.

Once we know that the dialectic between entropy and negentropy is the physical explanation of the reproduction of life, we can answer the question of how many original peoples survived during the past 10,000 years by agriculturally inhabiting their territories without predation of their surroundings. The response is because their ways of ecosystemic transformation, created from their diverse worldviews and life worlds, succeeded in adapting to negentropic processes of organizing living matter and reducing the dissipation of useful energy (Leff, 2004).

This is a good way of elucidating how traditional societies established their technical platform and inhabited their space coherently with negentropic self-organization. This includes the various agroecological techniques of using photosynthesis from solar energy, such as agroforestry management, the Mesoamerican *milpas* and *chinampas*, the Asian mountain rice terraces, the tropical systems of undergrowth, the Maghreb oases in the North African deserts and the Sahara, family vegetable gardens, or the agrosilvopastoral systems (Koohafkan & Altieri, 2010).

In all of these systems, the main principle is maintaining richness and biodiversity. Just to take the example of the Mayan *milpas* in Mexico, within the Lacandones' polyculture, some 51 species of plants can coexist in motley mosaics, 30 species of vegetables thrive among those of the Ch'ol and Tzeltal communities, 23 among the Tsotsiles, and 38 among the Yucatán Maya (Mariaca, 2010). These complex agro-systems are ecosystemic transformations that reveal how some indigenous communities have undertaken biomimicry, emulating many of the properties of natural ecosystems and how they have used negentropic biomass productivity to inhabit their territories in coexistence with biodiversity. They were achieved using interwoven techniques in harmony with the cyclicity of matter and useful energy.

These technical connections are place-specific and in tune with the precise conditions of inhabiting a territory that is different from all others and can only be understood by residing in it. Thus, in agroecology, there is no place for the monotonous repetition of universalized techniques; there is an infinitely open space for imagination and creativity regarding how to inhabit nature, connected to many community cosmogonies.

Again, the path of technical innovation is not closed to environmental thinking, just the opposite. We must loosen the reins on our imagination and cultural creativity so that social organization can be included in the negentropic process that makes up the natural order, such as the methodologies that Campesino a Campesino employs. This is precisely where the fundamental difference exists between agrotechnology and the agroecology technique, to be discussed in depth in the final section of this chapter.

Creativity: Technique and Technology

Techné is the key classical term for understanding the difference between the notions of "technique" and "technology." This difference is useful in parsing the role of agroecology in the context of the environmental crisis. In ancient Greek civilization, the word *techné* meant the skillful labor of the artisan. *Techné* expresses the ability of an artist to create something new. The concept used to mean creation and inventiveness for transformation through work. *Techné*, as the Greeks understood it, refers to the necessary imagination, intelligence, and initiative for creativity (Noguera & Bernal, 2013). It denotes a type of eminently practical knowledge that requires developing skills for inventing different strategies aimed at modifying nature.

The technique known as *techné* refers to the artist's skill in creating and innovating. Nonetheless, in a manner similar to the elderly artisan who has forsaken the creative group to which she/he belongs and ends up repeating older wisdom and actions (Duque, 1986), technology breaks away from its creative origins and become a simple, repetitive routine of activities by practitioners who use it (or are used by it). Technology begins with creative activity and ends in repetitive habits that are oriented at fulfilling the standardization required by the market economy.

As Armando Bartra suggests (2008, p. 80), the problem is not markets or meeting places where goods and services are exchanged, but rather that worldly activities are oriented by the whims of capital. In this "commercial absolutism," Bartra says, "the force of technological standardization" becomes predominant. This occurs because in capitalist output, equal goods are to be sold at equal prices, which is only possible when equal technologies are used. Thus, corn, soy, chicken, or sorghum producers, just to name a few examples, forced to act in accordance with the profit motive, are also obliged to incorporate the same technologies in a context of standardization and uniformity. The creativity characterized by the Greek concept *techné* is reduced to the universalization of set recipes that are tailored to the whims of capital reproduction.

Although in industry this logic has some meaning (at least in a version of economic rationality), when applied to agriculture, it is totally against nature and counter cultural. Agriculture, true agriculture, can only be carried out in a diverse world, where natural and cultural heterogeneity prevails. In Bartra's words (Ibid., p. 90), "The patterns of capital reproduction are essentially incompatible with those of natural human reproduction. This is so because the former seeks uniformity and the latter (although on closer examination they are just one), seeks systemic diversity." Nature's slow rhythmic cycle and the diverse cultures of its inhabitants, run up against the dizzying, intense, and monotonous technology of agribusiness. Oriented by the profit motive and price regulation practiced with industry, agro-technology exacts the standardization of human beings and nature (ibid).

Technology forgets about the creative gaze of the artist—who does not create his/her works by standardizing and repeating mechanical actions—and becomes a body of objective and generalized body of knowledge, whose purpose is related with the domination of human beings over nature (Ingold, 1990). In keeping with modern rationality, technology taps into our ability to reason to exercise control of the biosphere and steer a steady course for society toward a single path of "progress." Technology is linked to this notion and to the belief that if its development continues, we can all live better. This nineteenth-century optimism upon which we increasingly look askance hides the fact that capitalism and technology are inseparable, since only by investing in scientific-technical development can we continue to increase the value of large-scale capital. Yet technology is hardly apolitical, aseptic, and neutral. Since its modern origins, Western civilization began an unending technological race to increase productivity and, thus, never-ending capital accumulation (Bartra, 2008).

Agro-technology does not promote cultural creativity so that people can couple with nature through their technical know-how; rather it encourages users to follow manuals designed by technologists at the service of corporate capital. The triad of genetic selection, nutrition, and animal health in the poultry industry or the technological packages of genetically modified seeds that are resistant to herbicides and patented by companies are examples of how technology can be defined as a type of knowledge with some objective principles of "mechanical workings," according to Ingold (1990). Their validity is determined independently of the specific contexts of where they are applied in practice. The biotechnology of agribusiness is conceived ex situ and forces nature and the cultures that live in situ to become part of that productivist mindset.

In contrast, the *techné* in which peasant agroecology operates depends on context. It is specific to the place and requires a profound knowledge of the inhabited ecosystems and an imaginative skill to create and re-create multiple forms with an eye to their transformation. To cite but one example, the agriculture of the Andes has connected with the building of terraces and platforms, in keeping with climate and biotics of the Andean mountainsides. In addition to redefining spaces and coaxing on megadiverse habitats, vertical agricultural systems keep rain runoff from washing out the land, control landslides, and increase water absorption, thus improving soil humidity. In the highlands, Andean crop planters increase temperatures during frosts, and their canals "produce nitrogen-fixing algae that create a layer of organic earth and improve fertility when dry. They also help drain or save water, depending on crop requirements" (Altieri & Nicholls, 2000, p. 59).

Agroecological technique is a highly specific way of inhabiting a place. It is practical expertise, acquired through observation and imitation (Ingold, 1990). The agricultural knowledge of millennia exists, not because of formal verbal instruction transmitted generation after generation but because this knowledge is inseparable from action and practice (Ingold, 2000) in getting "one's hands dirty" by doing agriculture in specific ecological niches where people live. Agricultural novices learn by confronting situations and undertaking specific tasks (Ingold, 2000), such as predicting climate by gazing at the heavens and being attentive to animal indicators, such as bird or insect behavior. They also prepare furrows in keeping with weather predictions. They manage a wide variety of plant breeding activities according to short-term, mid-term, and long-term agricultural cycles. They learn about various techniques that vary by altitude and ecosystemic cropland niches. They also experience the complementarity of domestic animals in keeping with the needs of plants and atmospheric variables, as well as post-harvest techniques for processing and preserving products (Altieri & Nicholls, 2000). All of these practices are learned through pragmatic experience and encouraged by experienced agricultural producers who can teach the enormous creativity of ancestors in inhabiting living territories connected to environmental variables. Undoubtedly, communities that have lived for millennia in diverse ecosystems have instilled in each generation the technical traditions that come with modifications of these spaces through hands-on practice with environmental elements (Ingold, 2000).

Agro-technology, by contrast, is codified in formal instructions, does not pass from one generation to the next through practical and tacit knowledge as with agroecological techniques, but rather through a discourse that is codified in words or artificial symbols. In a very different way from agriculture, in which the specific techniques developed in situ are known with agro-capitalism, the "extension" of knowledge can be transmitted through formal teaching in settings that are outside sites of practical application (Ingold, 1990). With this type of knowledge, it is sufficient to faithfully adopt standardized procedures with no link to natural and cultural scenarios.

In other words, technology reduces the "technique" that is creative and specific to the place, to a type of de-contextualized knowledge focused simply on implementation. This technology does not reflect ingenuity, imaginative ability, and skill in knowing how to interpret nature's language and participate in its interior; rather, it is a set of recipes and rules that operate mechanically and whose implementation can be reproduced in any environment with no regard for the specific biocultural characteristics of the territories where executed. Yet, peoples' creative genius never unquestioningly accepts technology. There are always adjustments, modifications, and technical innovations made to technology developed by hegemonic scientific knowledge. Still, in most cases, agro-technology begins and is reproduced separately from the practical experience of people's cultures and their ecological settings. The agricultural producers who acquire them become dominated and alienated by a techno-power when they turn into simple mechanical workers.

Clearly, laboratory-invented eco-technologies, associated with negentropy and ecosystemic cyclicity, have an important role. The creativity of scientific knowledge should never be restricted. It should be encouraged within a "dialog of wisdoms," such that they do not end up using the agricultural producer. Ex situ inventions, created in research centers in a framework of an "ecology of wisdom" (Santos, 2009), should stimulate imagination and flexibility so that people can adapt them to their bioculturality and autonomy.

It is necessary to insist once again that post-development depends on a social order that connects structurally to coevolutionary, autopoietic, and thermodynamic conditions that make the miracle of life possible. Rebuilding techniques means giving them new meaning, based on the earth that is part of us and the richness of knowledge originating in cultural diversity, such that we may once again inhabit a world or, better yet, the many worlds that we have ceased to inhabit. This basis sustains agroecology and sets it apart from agroextractive business: the co-configuration of a profound sustenance that creates meaning and leads us to re-encounter ourselves and reconcile ourselves with life.

Chapter 7
The Future, Behind Us

Oh, such petty humanity
Popular saying

After all that has been said, it may seem redundant to go on about the industrialized nature of agribusiness, but I believe it is necessary to detail the characteristics inherent to the model of civilization based on industry, as well as the consequences of this mode of production. What makes agriculture "industrial" is a *rationality* based on a mind-set of bigger is better and the increasing consumption of material and energy. *Industrial agriculture* is a form of accumulation based on mass-producing food and raw materials, with increasingly complex technology, enormous installations, convoluted distribution systems, and the increasing monopolization by ever fewer multinational corporations of pre- and post-agricultural production. It is based on a system that produces cheap food to supply urban workers, creators of surplus value, and supply inputs to other industries in value-adding chains.

Industrial agriculture is also defined by extractivism. In other words, it treats nature not as a point of return, only as a point of departure in a chain of material production (O'Connor, 2001). It is a fossil-fuel-based agriculture, completely dependent on oil that functions through increased consumption of energy and minerals. The model also operates by treating humans as mere cogs in an industrial megamachine, which increasingly sentences agricultural producers to becoming workers, while the rest of humanity is reduced to being consumers in a rat race that changes men and women into so many "human resources" at the service of the system of exploitation. In Illich's opinion, industrialized agriculture is nothing more than a subsystem within a major hyper-technological system that increasingly decouples the countryside from the city, increases the disassociation between culture and nature, leading the progressively more urbanized human population to view "nature" as external, as people realize they are surrounded by a world that is increasingly de-natured and artificial. Indeed, the fundamental characteristics that define the marriage between industry and

O. F. Giraldo, *Political Ecology of Agriculture*,
https://doi.org/10.1007/978-3-030-11824-2_7

agriculture are extractivism, the giant-size mentality, and the division of labor among agricultural producers and consumers—which keeps the cities working by exploiting the countryside through parasitic relationships (Braudel, 1986).

These elements are the basis for an enormity that cannot help growing, moving fossil-fuel-based civilization toward an inevitable collapse (Tainter, 1988). Given its consubstantial counter-productivity, as mentioned by Illich (2006b), this rationality has long gone beyond its limits, doing the opposite of what it itself proposed: Instead of feeding, it creates hunger and malnutrition; instead of creating jobs, it replaces rural producers with machinery; instead of increasing efficiency and productivity, it inevitable reduces output. Most importantly, an intrinsic contradiction of the system makes it impossible to continue down the same path: It is physically impossible to keep an economy growing by feeding on a finite natural world (Leff, 2008). All of capitalist-industrial civilization was built on a foundation of abundant, available, concentrated, cheap, and easy-to-transport energy, i.e., fossil fuels such as carbon, gas, and petroleum, in addition to mining. These energy sources were stored in geologic strata for millions of years and, following excessive extraction, is now running out (Fernández & González, 2014).

As Ramón Fernández and Luis González (2014) noted, conventional oil, the easiest to extract and the best quality, has already run dry. Its zenith of extraction occurred in 2005, 2 years before the world's urban population surpassed that of the countryside. Other types of oil remain, some of which are nonconventional but of lesser quality and more difficult to extract in technical, financial, and energy terms. The zenith of all types of oil is calculated to occur in 2030, the same year as the pinnacle of gas; a decade later, coal will reach its peak extraction point. All of these fossil-fuel peaking points are interrelated, because lately the depletion of oil is having an impact on the other two types of fossil fuels and on possible alternatives for renewable energy, including wind, sun, geothermic, and tidal wave energy. Fernández and González maintain that, counter to mainstream thought, no alternative source of energy or any combination of energies can substitute oil. No alternative energy source, renewable or not, has oil's favorable characteristics: high energy density, multiple uses, easy to transport and store, and permanently available. Any possible transition to alternative sources will mean an increase in oil consumption, now running low, in order to extract large quantities of minerals, which also are at, or rapidly approaching, their peak extraction point. Neither biofuels, hydrogen, nuclear fission, nonconventional petroleum, nor a combination of these is a viable option for continuing to sustaining our civilization.[1]

[1]All renewable sources of energy are irregular and intermittent, need larger facilities, have storage problems, require mining, are mostly used for electricity, have very low energy efficiency, and are highly dependent on petroleum, in addition to the fact that they require extensive amounts of land, enormous investment, and petroleum to shift from fossil fuels to renewable energy sources. Other possible alternatives are unviable because agrofuels, such as bioethanol or biodiesel, have low efficiency, require petroleum, and compete with food grown for human consumption. Hydrogen is an energy vector, not a source of energy. Nonconventional petroleum has low energy density and low efficiency, depends on other resources for extraction (thus increasing costs), and has a large environmental footprint. Nuclear fission is not viable either because peak uranium extraction

We are at the end of the fossil-fuel era, and so industrialized civilization will need to make increasingly greater efforts to obtain energy (Fernández & González, 2014). In a context of global scarcity, accumulation by dispossession will increase, as will violence for control of the few sources of materials and energy that remain on this beautiful but despoiled planet. Perhaps Immanuel Wallerstein (2001) was right when he said that we are in a long-term structural decline of capitalism, which may be due to the difficulty of continuing to channel energy to the dynamics of accumulation. The rise of capitalism was based on natural resources that are running low, and this may be the key factor behind the collapse of the urban-industrial-world system. *Industrial agriculture* as a subsystem of a system in crisis also lacks alternatives. It is dependent on petroleum and phosphorus—an increasingly inaccessible mineral—and will have to deal with an energy restriction as it faces soil degradation, water contamination, air pollution, and the effects of climate change.[2] There are those who place greater hope in technology and the immense flexibility of a system to undertake technical adjustments and adaptations in order to stay the course. They may not notice at first glance that technology cannot create rocks or geological sediments in order to transform energy and extract materials, which is the root of the problem of an extractivist society.

Indeed, industrial society has difficulty acknowledging this dramatic situation. As Günther Anders said (2001),[3] this inability has to do with our cognitive limits to conceive the totality of such a large system. Once the scale of something goes beyond a certain limit, our ability to visualize all processes and what is at play diminishes. Anders holds that one of the characteristics of this industrial civilization is that our participation in this large-scale system seems isolated, given that our contribution is like that a small cog in a monstrous apparatus that we cannot comprehend. We are so fixated in such small segments of the global process of value added and consumption that we have difficulty perceiving the global magnitude of where we are immersed and the dimension of its consequences. Not only do we lose the ability to see the overall picture, Anders says, we also lose interest in understanding how the system works as a whole and its ultimate effects. The process gets worse, says Pierre Madelin (2016), when the consequences are invisible, such as those of climate change or nuclear radiation, i.e., invisible physical properties that go beyond ordinary perception.

The destructive machine in which we are situated is so huge and we are so detached from the ecosystemic relations of which we are merely users that we are blinded to the impacts that our actions cause. In a globalized and industrialized world that has reached these dimensions, we are so detached from the process as a

occurred in 2015, it has low efficiency, it depends on petroleum, or it can only be used to generate electricity (Fernández & González, 2014).

[2] I recognize my colleague Fabien Charbonnier for having thought about the urgency of planning for a future of agriculture without oil and with the effects of climate change.

[3] Pierre Madelin presented these insightful ideas during the Ph.D. seminar *Agroecology and Society* at the College of the Southern Border. See Madelin (2016).

whole that we cannot perceive its consequences or be affected by the repercussions of our actions as participants of this mega-system. We fail to see the impact of a plate served at our table prepared with food bought at a supermarket. We are not interested, and, in fact, we are unable to perceive how much water was used, carbon dioxide was emitted, and global pollution was generated during the growing, harvesting, processing, packaging, and marketing of each of the food items sitting on our plate. Or how many more resources went into making and distributing the necessary chemical inputs. Or how many others were consumed during the support-services segment so that those food items could be placed at our table. What Günther Anders says is true: There is no way to conceive the enormity of it all.

In reality, we have been tossed aside; we have awoken at the bottom of this industrial system that makes us consume and produce in a certain way, which only the large-scale industries can satisfy (Illich, 2006b). For example, in the big cities, people are forced to buy processed—and often genetically modified—food at supermarkets because nearby and local markets are disappearing in light of the emergence of big-box stores. At times agricultural producers are inserted in the system and dispossessed of their vernacular wisdom, or they find no other alternative than to increase the dose of purchased inputs. Thus, freedom is mutilated, reduced to a freedom to choose between one brand and the other, between one poison and the next (Marcuse, 1986). Yet, our frame of meaning is circumscribed in its interior, and we resist making changes in our thought patterns because, in a way, like Janus, we are only shown the seductive face of comfort, while on the other side, the devastation of the land, the exploitation of fellow humans, and the misery etched on the face of the wealthy few are hidden (Noguera & Giraldo, 2017).

Enthralled with luxury, we have fetishized the images of progress and hyper-technologization in the world. Our expectations about the future are like Hollywood movie "futures," i.e., increasingly artificial, hyper-technologized, gray, treeless societies. Our imaginary is the dystopia of the technological fetish, and, like in *The Jetsons*, urban life invades the world; agricultural producers are replaced by electronics, robotics, nanotechnology, and satellites; and we feed on pills made for astronauts. In truth, these are linear scenarios based on the anachronist idea of scientific-technical progress, in which, paradoxically, we do not seem happier, just more alienated and transformed into cyborgs.

Yet the images of the future projected by the collapse of energy sources and the inevitable end of industrialized capitalism are not those of Hollywood. Some, undoubtedly, are apocalyptic, since they increase the risk of progressively destructive wars aimed at eliminating excess population and controlling territories that contain the last bastions of fossil fuels and sources of water. Others images, however, portray distinct possibilities: small-scale, deindustrialized, and de-urbanized societies, which find new meaning in the commons, proportionality between intellectual and physical work, with simpler, repairable, and long-lasting technologies, and reflect new ecological balance emerging from the ashes of the previous industrialized society (Burkhart, 2012). Undoubtedly, this is a utopia, an excellent reflection of which is Cuba and its agroecological territorialization after the fall of the socialist bloc and its compulsory "de-petrolization." However, many other future possibilities exist when we forsake linear and teleological ideas of progress and envision possibilities within discontinuous scenarios. The specter of growing artificiality can appear in a linear context of logical development "that points to urban societies as the ultimate goal" (De Landa, 2011, p. 13), but not when we conceive of a critical, disruptive scenario where the system has to self-organize differently.

I would like to delve into this last idea with the help of Anabel Ford and Ronald Nigh's (2016) magnificent book *The Maya Forest Garden* about the collapse of the Mayan civilization.

Possibilities of Worlds Turning Green Again

> That which still does not exist is walking behind us, and we can't see it.
> Abelino Dagua, Misael Aranda, and Luis Guillermo Vasco

Researchers Ford and Nigh have dedicated their lives to studying the Mayan people. Ford is recognized for her archeological discovery of the ancient center of the Mayan city of El Pilar on the Belize-Guatemala border. Nigh is a prominent anthropologist who has studied the *milpa* for close to 40 years. These two researchers, informed by paleo-ecological, archeological, and ethnographic records, reject the conventional hypothesis that the Mayan civilization ended with the disappearance and destruction of their environment. The mainstream account posits that the collapse of Mayan civilization was due to overpopulation, drought, and the ensuing deforestation and degradation of soils that resulted from transforming the jungle into croplands. According to Ford and Nigh (2016), the first error of this narrative is to conceive of the Maya as a disappeared people, as those of us who live in this lovely part of the world can attest. The second error is to think of the Maya from a Western perspective in which forests and agriculture are incompatible. This idea ignores the coexistence of crops and forest in the Maya's agroforest systems and their sophisticated way of inhabiting the jungle, whose history dates back some 8000 years.

We need to go back in time 2000 years to tell the complete story. The first inhabitants arrived in this region of the planet 10,000 years ago. At that time, the area was quite different from the current humid tropical jungle; it was rather more arid, with a dry and temperate climate. This means that the Maya were present as the area became a jungle. In fact, it is widely accepted that the jungle in the present-day Petén of Guatemala and the Yucatán and Lacandon Jungle of Mexico is not a pristine territory but rather a human co-creation. This biocultural coevolution was possible due in great part to the Maya's profound knowledge of fire management and the succession of forests. For 8000 to 4000 years ago, the Maya developed a system of cyclical polyculture based on corn, known as *milpa*, still used among their descendants. The lowlands *milpa* has a cycle that begins with the slash-and-burn method of clearing jungle spaces, which is then planted with annual polyculture for

4 years until corn production begins to diminish. This is followed by 16 years of natural regeneration of the forest, after which a new cycle resumes as planting occurs in the same place. During this 20-year cycle, the fauna feeds on numerous successional stages, while rural producers put the plants to use in different ways. Ford and Nigh's thesis holds that, because of a long period of climate stability, the Maya developed a system of *acahuales* [fragmented forests] that both created one of the most biodiverse jungles in the world and set the stage to sustain the earliest settlements and encourage urban growth of the Classic Maya.

Ford and Nigh's explanation for the change that led to Mayan urban societies posits that when the lengthy period of climate stability came to an end, a period of instability began with unpredictable rainfall and climate extremes. The climate chaos that occurred in the pre-Classic period created conditions that led dispersed and nomadic peoples to gather in permanent settlements in well-drained areas with a stable water supply. As the climate grew dryer, water scarcity drew agricultural producers to urban centers. Gradually, the *milpa* system intensified, which permitted higher population density that reached its peak between 1400 and 1100 years ago with some 1000 inhabitants per square kilometer. This situation was largely enabled by the flow of high-intensity energy provided by the Mayan forest garden and the *milpa* system during a period of renewed climate stability. The Maya did not increase their population as they occupied the jungle. A more accurate depiction is that during the Classic period, the population lived in cities with higher population density, in coexistence with a forested landscape of high-yield *milpas*.

What, then, caused the collapse? Ford and Nigh's argument is compelling. It was neither deforestation nor overpopulation. These alternatives are rather more a reflection of our own time in which monocrop agriculture cannot coexist with forests and the lifestyle of an industrialized society is incompatible with the time and principles of ecosystem regeneration. What actually occurred is completely different: A collapse of the sociopolitical system in which elite traditions were unsustainable and rulers lost legitimacy due to their warfare follies. In this scenario, Classic Mayan leaders' power to collect taxes was weakened, and the population retreated to the jungle, abandoning the impressive urban centers. The change in settlement patterns did not occur overnight. It happened over many decades, even centuries, and was never totally carried out, given that a significant number of residents continued to live in the post-Classic centers when the first Spanish conquistadors arrived years later. The desertion of the monumental buildings can be explained by the failure of a political and economic system. Yet this says nothing of the fate of agricultural producers who, in fact, increased their numbers until the start of the Spanish conquest. What occurred was a transformation in the ways the Maya inhabited the ecosystem, by which the lower classes, now free from the responsibility of sustaining great cities, reorganized into smaller and more dispersed units.

I discuss this version of events to back the point I made previously, i.e., civilizations do not advance in a linear manner. Indeed, Mayan history shows us that their urban settlements were only an intermediate lapse between two mainly rural periods. A 4000-year-long agricultural phase preceded an urban phase that lasted

3000 years, followed by a mostly rural phase that has lasted 800 years. The fact that history does not advance in progressive stages, nor is it teleological, nor is necessarily headed for an increasingly technologized lifestyle is a clearly taught lesson from the Mayan people, recognized for their knowledge of astronomy, mathematics, and impressive art and architecture. Conditions did not allow them to continue "developing" their astronomical or mathematical knowledge progressively. The history of Mayan civilization, and many others, shows us that progress is not always linear. There are often breaks, twists, and events that lead to dramatic changes. History may seem to be a spiral, since we never return to the previous point; it is more likely that contingent events will occur that leave us in situations that are similar to those faced by our ancestors.

The point I want to make is that our future does not have to look like a Hollywood movie. There are other possibilities, such as a massive outflux to rural settlements, so that cities, far from disappearing, will coexist as smaller mosaics within agroecological habitats. What conventional wisdom says about Mayan civilization reflects the fears and fetishes of our Western present (agriculture that is incompatible with forests, overpopulation, and ecocidal societies that degrade the sustenance they depend on), and the specter of increasing artificialness, and less about our future. If we think seriously about the difficulty of continuing this suicidal civilization, we can begin to imagine other "greener" futures. The collapse of civilization is not necessarily our destiny. It is an opportunity for us to undertake changes and adapt to new circumstances. Although it is true that the sources of energy and materials that sustained the growth of industrial civilization and capitalism are running low, we should begin to prepare for post-extractivism and energy autonomy, whereby knowledgeable people learn to transform energy from local sources without maximizing entropy, and we become part of an autopoietic cyclicity that makes life possible.

The phenomenon of "new rural producers" is perhaps a preview of the changes that await us. This may sound like speculation, like any prediction, but based on what has been discussed throughout this book, I do not think it is absurd to believe that this may be a good solution to avoid the morass in which we currently find ourselves. Like Leon Tolstoy, Gandhi, the counterculture movement of the 1970s, or the eco-villages that are created in a growing number of locales, the people who are abandoning the cities also have an honest desire to flee the bourgeois dream. Their actions break with the illusions that sustain the system. In Lacanian terms, they stop embodying the desire desired by others. They refuse to identify with the capitalist symbolization that structures the lives of those of us who remain within the system's orbit. It may seem naïve, or romantic, to eschew scientific-technical progress that capitalist modernity uses to chart its future. On the contrary, I believe that it is naïve or romantic to continue to contemplate more growth, development, extraction, urbanization, and technical sophistication designed to manipulate our lives. The Mayan experience demonstrates that when an urban settlement is not working, the urge to abandon the metropolis grows stronger. Thus, today it is not more pertinent to query if rural inhabitants will survive the forces of capitalism, or what role they will play in the revolution—Marxist questions from the twentieth century regarding the rural population. It is more timely and useful to study how the rural sector can be repopulated through agroecology, a process that could occur in the context of an imminent collapse of industrialized society.

It is not my intention to predict the future. We know not when the collapse will occur, and it is probably not the moment to dust off our crystal ball to foretell the future. The idea is not for us to merge with the horizon, as the dogma of progress would like, but rather take hold forcefully of our present and review the possibilities that make sense, here and now, irrespectively of what happens in the future (Esteva, 2016). We are probably at the end of a historic cycle, but rather than wondering how and when the collapse will occur, it is best to shine more light on how, in the midst of this chaos, alternatives can take root, without losing sight that the old status quo is disappearing (Marx & Engels, 2001). I refer to the fact that new permaculture producers, the rural inhabitants that implement agroecology and share its benefits with neighbors, as well as the many initiatives in economies of solidarity that increasingly arise, may be part of an embryonic civilizational transformation (Marañon, 2013). Everyday people, here and now, are creating post-development practices, as we saw when we discussed the peasant-to-peasant methodology. Theirs are not minor rehearsals; they are creating the key aspects needed to replace obsolete capitalist society.[4]

Ford and Nigh's understanding of the Mayan collapse shows how its agroforestry systems were important insofar as they were the basis that permitted both the change in settlements from forest to urban life and the opposite shift later. Actually, over thousands of years, the Maya's greatest wealth was and continues to be their in-depth agricultural knowledge that has allowed them to inhabit within forests or, as the theory proposed in the previous chapter states, the satisfactory connection and interaction between culture and environment facilitated the system's integrity over almost eight millennia. What is crucial is our understanding of the importance of what today we call agroecology for the transition of the civilization in which we are immersed and for imagining a distinct landscape from the currently prevailing one that is decoupling society from nature.

To understand the important role of agroecology in the transition of civilizations, I want to return to Augusto Ángel-Maya's approach to the environmental problem. According to Maya, a Colombian philosopher, the conflict in our time stems from

---

[4] As Gustavo Esteva (2009, p. 5) lucidly stated, "The following metaphor may reflect what is currently happening. We are all in a boat in the midst of a perfect storm. Within the engine room, politicians, scientists, social leaders, government workers, political parties are having an intense argument… Everyone has an idea about how to address the problem. They are so wrapped up in the debate that they don't realize that the boat is sinking. But the people on deck are vividly aware. Some, taking matters into their own hands, jump from the boat and drown. Others organize and, dividing into small groups, begin to build rowboats and rafts and drift away from the boat. The people find ways to coordinate their efforts further and suddenly realize they are in the middle of an archipelago of conviviality. Off in the distance they see their supposed 'leaders' sinking with the boat."

an incompatibility between ecosystems and culture, because the former has its own statutes, which do not coincide with the cultural realm. This in no way implies that human beings are forbidden from undertaking changes to their surroundings, nor are they forced to keep virgin nature intact, like some showcase display that can be admired but not inhabited. Quite the opposite; we are a species that made ourselves into what we are by transforming ecological niches. This is why, according to Ángel-Maya (1996, p. 52), "the environmental problem is not one of 'preserving' nature, but rather of modifying it well, even though it might be necessary to 'preserve' it in order for proper conservation to be achieved." A good example is the ecosystemic transformation undertaken by the Mayan civilization, which was able to inhabit the forest by modifying it, co-creating it, and transforming it into a forest garden.

Nonetheless, the conservationist approach to sustainable development does not contemplate inhabitants within ecosystems and actually advocates separation, since it considers that the presence of humans is harmful to natural systems (Ángel, 2003). From this break in tradition thinking, islands of strategic, protected ecosystems emerge surrounded by oceans of devastation. In one corner, we have human beings and all the privileges accorded to them and, in the other corner, natural protected areas in which humans are not allowed to do almost anything. This is undoubtedly a dualist structure that reflects the crisis of our civilization, such that human cultures are treated as a nuisance that invade nature, even though, paradoxically, this line of thinking covertly accepts capitalist expansion and onslaught against nature in "non-strategic" corridors.

The landscapes of development exemplify with extraordinary clarity the institutional practices that Giorgio Agamben (2017, p. 155) called "death policies." This Italian philosopher wrote, "There is a line that signals the point where a decision about life becomes a decision about death and where bio-policies can thus become death policies." I believe that the line indicated by Agamben can be found in the legal frameworks that regulate sustainable "land-use planning" policies, whose content stipulates where life can exist and where the borders are, beyond which "projects of death" are permitted. In conservationist policies, we must consider, just like the authorities who regulate natural protected areas, not so much what is included within the areas to be protected, but what is excluded from protection. Without these policies saying so openly, their tacit implementation means that a portion of territory is to be placed outside the legal framework in order to be preserved. Thus, life on the one hand and death on the other become political concepts that acquire meaning as the result of a governmental decision.

From a different angle, Ivett Perfecto and John Vandermeer (2010) call the division that separates biodiversity conservation from production "land sparing." According to this dualist way of understanding environmental problems, ecosystems and culture are treated as independent spheres, which in agriculture is highlighted when institutional and multilateral policies defend "sustainable intensification," in order to have the highest productivity possible in one area and free other lands for conservation. Perfecto and Vandermmer question this stance and, in contrast, envision something closer to the landscapes of the Mayan civilization, which they call "land sharing." These, in contrast to the landscapes arising from the sustainability schism, have integrated society and ecosystems. Perfecto and Vandermmer's argument posits that organisms that are in fragmented landscapes suffer local extinctions, while a diverse agroecological matrix encourages, much like a bridge, their movement or migration between patches of their natural habitat. Within "oceans of devastation," population movement is repressed, they say. Yet a combination of fragments made up of natural vegetation that are embedded in an agroecological matrix can produce another vision of what might be a changing but integrationist landscape in which human beings couple with the ecosystemic order.

Inhabiting diversified agro-landscapes, entangled with mature forests, would mean relating to agroecosystems as ecosystems and not as food factories; this would imply viewing the system "holistically," where its components interact in different ways. One form of biomimicry would endeavor to make a system as diversified as possible by imitating natural regeneration, making the system appear similar to the latter stage of ecological succession. In this way, as we explained in the previous chapter, synergic relations would increase, redundancy would be encouraged, and complementarities would be stimulated, thus increasing the paths for the flow of nutrients and energy. This would, of course, make the system more resilient and lessen artificial entrances and exits by using generated waste as inputs, without producing further waste (Gliessman, 1998; Grifon, 2012). This is, undoubtedly, a post-extractivist utopia in which—like the landscapes described by Ford and Nigh—it is possible to conceive of forests integrated into a high-quality agroecological matrix with agricultural producers living in its interior.

From a political point of view, Illich reminds us, the agro-landscapes of a civilizational transition would place a technological ceiling on the means of production, not just because of the physical impossibility of sustaining an industrialized system but also because monopolistic and hierarchical domination makes it impossible for society to be in control. Having paid the enormous cost of alienation and destruction, we have learned that the more sophisticated technical apparatuses are, the less control society has over them.

We repeat: We are not marching inevitably toward a megalopolis as the final goal of history. Instead of the unidirectionality of progress, we are called on to undertake a radical break, forsake this relationship of domination, and recognize that as human beings we belong to the earth in a tightly woven relationship. I believe that agroecology is an integral part of the transitions to post-extractivism, i.e., transitions that require not only a change in the technical platform but also a profound ontological and spiritual change, so that we cease conceiving of nature as an overflowing warehouse of available resources to satiate the needs of an increasingly industrialized society.

Ontological Transformations, Spiritual Transformations

Extractivism means much more that gigantic quantities of materials removed to feed industrial society (Gudynas, 2009). It means that nature is unhidden and forced into the spotlight as merchandize. The path opened by Parmenides (his philosophy's reluctance to conceive of darkness, shelter, and rest), is today trod by the unfettered growth of supply, the perpetual gains in productivity, and the calculated devastation, all of which leave footprints in the deserts of development. Heraclitus insisted that *nature likes to hide itself*. Today we understand the deep meaning of those words: Hiding is a form of rest so that life can bloom again. If we forget this principle, we can succumb to exploitation, administering dispossession, and inhibiting the self-generation of vital connections. Understanding Heraclitus's path means establishing a very different type of relationship than that the one touted by Parmenides, i.e., a more cautious, respectful, wise relationship. Instead of dominating, manipulating, and voraciously pumping out hydrocarbons from geological depths, of intervening in nucleic acid through synthetic biology, genomic inhibition, or transgenesis, we must adapt human life to ecosystemic cycles. As a way to take the path of Heraclitus, ecological agriculture, instead of objectifying Earth and exposing her as is done by industrialized extractivism, endeavors to understand the agroecosystem profoundly, recognizing with wisdom, and with a bit of humility, the many ways that relationships and interactions occur within the inhabited place.

In addition to a more caring relationship, ecological agriculture has profound ontological implications that I wish to discuss further. As we have iterated, human beings do not just modify places. Places also modify us. Thus, this type of agriculture is not simply a means of production. It is also a very particular way of being affected. Just as landscapes that arise from disciplining and mutilating the earth produce ways of "being" associated with the maps of agroextractivism, the landscapes of diversity, multiplicity, and the blooming of life also produce ways of "being"; these are affectations that occur on the bodies that inhabit these spaces. We as human beings empathize and connect with places, because far from just inhabiting them, they are spaces that inhabit us. Clearly, we need a profound ontological—and spiritual—transformation, in which, instead of feeling isolated, absorbed in our own and perfectly staked-out "me-ness," we ought to understand that we are interdependent and hyper-interrelated beings. Zen Buddhist monk Thich Nhat Hanh (1998) called this the "inter-being," i.e., an understanding that we Are not, but rather we "Inter-Are," with the water we drink, the soil we walk on, the air we breathe. Given these unsettling times, the ontological and spiritual change that is so urgent cannot be understood in isolation from the conditions where we establish our residence.

Furthermore, if we live in increasingly artificial cities, with greater amounts of cement and steel, more intelligent buildings, and greater amounts of electronics and robotics, these spaces have a great impact on our ways of "being." Similarly, if we live among motely landscapes with polycultures interspersed in forests, these spaces will largely provide the relational, symbiotic, and reciprocal meanings that we greatly need. My point is that no matter how our transformations affect ecosystems, they will have an impact on how we understand ourselves. The modification of our self-understanding as beings separated from nature, isolated from our fellow human beings, and beholden to the wants produced by the capitalist machine, for a different self-understanding, more closely aligned with an ontology of the "inter-being," would be facilitated if we change the ways we transform the ecosystem. We cannot transform our "being" if we do not implement ecosystemic modifications based on total respect for the vital relationships that inhabit us and which we inhabit.

As cultural beings, we are unable to adapt to a pre-given environment, just as we cannot simply occupy a niche within an ecosystem. We invent habitats and we inhabit their interior. But these transformed spaces penetrate the core of our bodies, affect us, they shape us as inhabitants (Pardo, 1991). We discussed this when we analyzed the landscapes of extractive agribusiness; but we should also consider this aspect in a context of agroecological landscapes by investigating how the imprints, the footprints created through the art of ecological agriculture, literally "happen" to those that inhabit reshaped territories in spaces that have been transformed. This is a fundamental matter, given that inhabiting random hyper-technologized worlds is not the same as worlds that have become green once again, such as those of an agro-ecological utopia. Each scenario creates a different type of inhabitant. In the former, the closed circuits of disconnection are reinforced. In the latter, we free ourselves from a metabolic rupture, and we regain awareness of our impact on the world, eroding the tragedy of that industrial enormity, which we are unable to conceive in the interior of this system.

The decisive aspect consists of understanding that the ontological and spiritual changes are profoundly pragmatic, given that they do not occur in a deterritorialized manner. They take shape in the places where we are inhabited. It is in the "being-here" where we build the symbolic network through which we interact with nature. If our lived experience occurs somewhere else, it is there where the imaginaries and the cultural signifiers that shape our behavior will emerge. Thus, if our everyday habitation occurs in a diversified landscape, as suggested by the agroecological civilizing project, the perception and symbols will be incoherent with the dualisms of modernity. Symbols are not constructed in a manner that is different from the way societies experience the world. They arise in relative coherence with the characteristics of the spaces where we establish our residence. True, "we cannot suggest structural solutions to our civilizational crisis without profound modifications in the cultural symbols of that same civilization," as we discussed in a previous publication (Giraldo, 2014, p. 205). Still, the harmonization of symbols and values associated with ecological cycles depends on the dialectic that can occur during ecosystemic modifications.

These musings are based not on some phenomenological abstraction but on empirical evidence. In Cuba, for example, a large-scale agroecological reconversion was implemented not because everyone was convinced of the need to undertake ecological practices in agriculture, but because a crisis forced the Cuban people to transition. Nonetheless, during the shift to agroecology, "environmental and social criticism of the Green Revolution and its impacts grew, while environmental values flourished" (Machín, Roque, Ávila, & Rosset, 2010, p. 32), in the words of some participants. Clearly, this is a good way to demonstrate that "we are what we do," that there is a coincidence between "doing" and "being" (Maturana & Varela, 2003), and that meaning is not resistant to change. During times of tension and crisis, history teaches us that cultures are forced to make changes to adapt to new circumstances and it is a practical response to symbolic metamorphoses (Wolf, 2001). When societies go beyond their limits and alter the conditions of their own inhabitability, "environmental impacts bring pressure on the culture system forcing it to change or disappear" (Ángel, 1996, p. 95).

The principal problem of this new millennium is learning to live with others, to inhabit alongside the human being with whom we are "inter-being." In this enormous challenge of "inhabiting the difference that inhabits us," as José Luis Pardo says (1991, p. 144), we must learn to make changes that are in tune with the language of nature. We will need to keep creating habitats without altering biotic equilibria or transgressing plants and animals' natural niches and their symbiotic relationships. Nature does not cry out "don't touch me," but rather it calls for a transition to an ecosystemic order that is in harmony with the ways of "being" based on interrelationships, interdependencies, and complementarities. Ecological agriculture is an answer to that call, given the physical impossibility of remaining in the midst of this suicidal odyssey, but also because living with others and among others is not an option. This is a subject that is organic to our nature as inter-beings.

Technical, economic, and political changes do not occur out of thin air. They are accompanied by deep ontological changes. This means that these changes should not be understood independently, nor should we think of some changes being the consequence of others. All of these changes are bound together and form a reciprocal relationship. The point being that we should understand to what extent and under what conditions ecological agriculture should accompany ontological change and be part of the background, where cracks in structures of truth and the order established by institutional biopolitics of agricultural development arise, while the ontological changes that have produced barbarity flourish. Agroecology, as a science of this "relational doing," can well be a part of the knowledge of interdependencies that we would do well to learn fully, in order to inhabit the Earth that we have uninhabited. Nonetheless, if science is based on sustainable development discourses, it will not be up to the task of transformation such as the one required by the impending civilizational collapse we face. Agroecology must be understood as a constitutive part of a new civilizational matrix that recognizes that agriculture is an irreplaceable foundation for a geo-poetics of post-development and post-extractivism, i.e., of that to which we attach the prefix "post," because it still lacks a name, if we so deem to name it. For now, we shall label "civilizational transition" that slow abandonment of anthropocentrism, utilitarianism, individualism, and the entire scaffold of values that prop up this schizoid civilization, as we step onto a different path and begin to commune with nature's own wisdom. Beyond being experts in nature, we need to be wise about nature, agroecologically wise, as we learn to "undertake" geo-poetic gestures far from that branch of science that seeks to strip nature of its secrets.

Better still, we should become more familiar with a less-pretentious science that celebrates both uncovered truths and the mysteries of the hidden to which, as agriculture producers teach us, we need to pay Tribute.

Technically, we may still be able to undertake further genetic manipulations, invent sophisticated devices, and design machines that excavate ever deeper pits. Yet, wisdom, which is so lacking in science and of which modern technology is further bereft, reveals that sometimes it is best to refuse, abstain, do nothing, and instead open our minds to the magic enigma of life. Knowing how to preserve life in its hiding place needs astonishment, of that which we refuse to know with the tool of calculating reason, and open ourselves to the poetry that does not extract, does not uncover. Preserving life means, in contrast, contemplating how the fabric of life seeks to withdraw and rest in the profundities where darkness reigns, making possible the surprising brilliance of life (Noguera & Giraldo, 2017). If, as the Zapatistas teach us, other worlds are possible, another, very different science, a science-poetics, we might call it, can accompany those transformations that require profound spiritual changes. The crisis of civilization is a crisis of meaning, in that a science cannot keep from being part of the problem if it refuses to shift away from the direction it is heading with its calculating mind-set and reset its course toward a heartfelt science, which will forsake intervening, dominating, manipulating, and subjugating, as it focuses on understanding how our cultures can inhabit a world that we currently occupy as if we were professional armies.

The political ecology that we have discussed in this book has enabled us to study the politics of agriculture not according to a set of institutional practices, such as laws or legal regulatory frameworks, nor with the particular interpretative lens of political parties, politicians, or elections. Rather, we have approached it from an ontological dimension, i.e., with the understanding that every political project entails a particular way of meaningfully understanding and positioning ourselves in the world. By this, I mean that we should avoid a narrative of the morally superior "good ones," versus the "bad ones" who pillage the soil they depend on. There is no evil in the mainstream way of being-in-the-world; it is not even an ethical failure (Mardones, 2016). The industrial agroextractivism described herein is simply a manifestation of the cultural context set before a backdrop of Cartesian separation of nature from society, individual from community, mind from body, and subject from object. From this background, our self-understanding takes shape as reasoning subjects who inhabit a world of individuals separated from each other and in which the external to human beings is considered an inert object available to be manipulated and dominated. From such a mind-set based on separating nature and the individualization of fellow human beings, a metaphysical thinking and an economic rationality emerge to lay the symbolic foundations of contemporary agricultural capitalism. That is why the root of the problem is ontological. I have no doubt that the organizations that promote development are well intentioned. But based on a self-conception as separated "beings," they fill with meaning their domination of nature, their emphasis on surplus value, the egotistic foundations of liberal economics, the manipulation of the nonhuman, and competition.

Social movements that defend agroecology, inspired by practices of traditional agriculturalists, fisherfolk, nomadic pastoralists, and other popular sectors in the countryside, in addition to more recent practices, are touting a political project that may lead down a path to relational ontology. For many people, this is just common sense, insofar as daily life cannot be conceived in isolation from the territories of life and the communal relationships that undergird it (Escobar, 2015). In Latin America, the Living Well Project—*Buen Vivir*—a heuristic meeting space for relational ontologies that promote post-development (Gudynas, 2011), has tried building an inter-being politics, without opposing nature with something from another realm. This politics understands that we are part of a knot of relationships, that the fabric of our bodies is weaved by the roots of Mother Earth. Therefore, we cannot live well if other human and nonhuman beings live poorly. A geo-poetic-politics that, far from the "green" discourses of modern ontology, is not based on a body unconnected from the world that then seeks to encounter it. Rather it is one that from the start acknowledges that we are body-earth (Noguera, 2012), bodies rooted in vital weaves, corporal prolongations of other beings that inhabit us and which we inhabit, and that any transformation of the world begins with inhabiting alongside everything else in a way that is mutually enriching (Blaser, 2013).

The political ecology of agriculture is best understood if we consider these two projects (which here are much summarized and leave grey areas with nuances among them) as a scenario of ontological conflict arising from a discrepancy of principles and meanings that gives rise to each one of the disputed ontologies (Blaser, 2013). When we speak of conflict, we are referring to the center where political practices are possible. As Žižek says, there is no possibility of overcoming antagonism. In fact, it is both inevitable and unwise to conceive of a fanciful harmony in which the parties in conflict agree to a peace found only at gravesites. Acknowledging that inter-being implies accepting otherness, recognizing it in its radicalness, and accepting its presence. That is why we constantly need the presence of the contradictor, because without antagonism, not only would we not engage in politics, we would also be unable to nourish our project dialectically. The ontological conflict that emerges between antagonistic actors also implies the presence of dissention in the symbolic sphere, in the immaterial territory of ideas (Fernandes, 2008). This is how the largest movement in defense of peasant agroecology put it:

> Ours is the "model of life," of the countryside with peasants, of rural communities with families, of territories with trees and forests, mountains, lakes, rivers and coastlines, and is in firm opposition to the "model of death" of agribusiness, of farming without peasants or families, of industrial monocultures, of rural areas without trees, of green deserts and land poisoned by chemical pesticides and genetically modified organisms. We are actively challenging capital and agribusiness, disputing land and territory with them. (LVC, 2015b)

I believe that we can understand the political ecology of agriculture in its full dimension with this last quote from La Vía Campesina, which clearly sets out the ontological conflict that informs each party's arguments; thus we can infer what is at stake in this dispute. The role of politics is not one of ending the conflict between antagonistic actors, but rather to make it visible and, in Gramscian terms, allow the struggle for common sense in agriculture to take place. Agroecology acquires its full political status when it tries to weaken the modern structure of producing truth about the agri-food sector through persuasion, as agroecological social movements have done throughout the world. Nonetheless, this goal cannot be fully radical if it does not acknowledge that the political conflict is ontological, which means piercing through the symbolic order that undergirds the metaphysical meanings of agribusiness, and encouraging practices of inter-existing, leading to profoundly questioning the cultural project of modernity and development.

In this book, I endeavored to demonstrate here that the social processes behind agroecology are already building post-development and questioning through their practice the developmentalist objective of shaping people's lives by incorporating them into state structures and market practices. I hope that the book has shown that when mobilizing methodologies such as peasant-to-peasant are implemented, it is possible to carry out an autonomism of the place, keeping distance from institutional spaces where Western metaphysics is administered. We have slowly come to doubt that state structures are the best arenas for political dispute. This does not mean that the legal frameworks, institutions, and the bureaucratic scaffolding are nothing more than a large piece of property that must be occupied by "moving the fences" through conquests, as Peter Rosset has suggested. We must understand that people's destiny cannot depend on the "representatives" that hold institutional office, or relinquish power to outside control so that their lives can be governed. Agroecology and its social processes have taught us important lessons for post-development and for civilizational transitions, one of which is that grassroots organizations can take power in their hands without then handing it over to the state. Now, I do not want this critique of statism to be understood out of context, implying that I believe the state should cede its power to the market, as pundits of economic liberalism favor. Rather, power should be seized by the people, meaning that they must assume responsibilities for their own lives by practicing self-management, self-determination, and autonomy, aspects of which are already underway, and from which we have learned much. As long as the state, as we know it, continues to exist, it will always be the antagonist of social movements, which does not mean it should be ignored; rather, it should be controlled and made to obey, just as people should resist the temptation to become part of a state bureaucracy.

It is not a matter of patiently waiting for the collapse to end or for some mega-event to change the system overnight. Civilizational changes take a long time, compared to our own lives, even though when seen in a historical context, they occur rapidly. Without us knowing it, it is possible that agroecology has already become a part of the civilizational transition and people, experiencing radical changes in their territories—through social processes such as peasant to peasant—are making real ontological and spiritual transformations. These changes are necessary if we are to learn to listen to the source of wisdom itself, which lives in the surrounding water, fire, air, and land. What we are proposing is to connect, to listen, to empathize with the natural elements, where everything we need to know about the world resides. If we live with sympathy for the territory through agriculture, we are already a part of the transition toward post-development and post-extractivism, as we continue to look for ways to bring harmony and complementarity to the conditions that make life possible, by learning to connect to the wisdom of nature.

References

Addiscott, T. M. (2010). Entropy, non-linearity and hierarchy in ecosystems. *Geoderma, 160*(1), 57–63.

Agamben, G. (2017). *Homo sacer. El poder soberano y la nuda vida I*. San Cristóbal de las Casas, Mexico: CIDECI-UNITIERRA.

Alimonda, H. (2011). La colonialidad de la naturaleza. Una aproximación a la Ecología Política Latinoamericana. In H. Alimonda (Ed.), *La naturaleza colonizada Ecología política y minería en América Latina*. Buenos Aires, Argentina: Ediciones Ciccus-CLACSO.

Altieri, M., & Nicholls, C. (2000). *Agroecología: teoría y práctica para una agricultura sustentable*. Mexico: PNUMA.

Altieri, M., & Toledo, V. M. (2011). The agroecological revolution of Latin America: Rescuing nature, securing food sovereignty and empowering peasants. *The Journal of Peasant Studies, 38*(3), 587–612.

Álvarez, C. F., & Velásquez, H. I. (2013). Exergía en sistemas biológicos: Aproximación holística para el estudio de ecosistemas y el manejo ambiental. *Producción+ Limpia, 8*(2), 106–127.

Anders, G. (2001). *Nosotros, los hijos de Eichmann: carta abierta a Klaus Eichmann*. Barcelona, Spain: Paidós.

Ángel, A. (1995). *La fragilidad ambiental de la cultura*. Bogotá, Colombia: Universidad Nacional de Colombia-Instituto de Estudios Ambientales.

Ángel, A. (1996). *El reto de la vida. Ecosistema y cultura*. Bogotá, Colombia: Ecofondo.

Ángel, A. (2002). *El retorno de Ícaro. La razón de la vida. Muerte y vida de la filosofía. Una propuesta ambiental*. Bogotá, D.C., Colombia: PNUD-PNUMA-Instituto de Estudios Ambientales.

Ángel, A. (2003). *La diosa Némesis: Desarrollo Sostenible o Cambio Cultural*. Cali, Colombia: Corporación Universitaria Autónoma de Occidente.

Anino, P., & Mercante, E. (2009). Renta agraria y desarrollo capitalista en Argentina. *Lucha de Clases, 9*, 69–110.

Anseeuw, W., Boche, M., Breu, T., Giger, M., Lay, J., Messerli, P., & Nolte, K. (2012). *Transnational land deals for agriculture in the global south. Analytical report based on the Land Matrix database*. Bern, Switzerland/Montpellier, France/Hamburg, Germany: CDE/CIRAD/GIGA.

Bajoit, G. (2009). La Tiranía del gran "ISA". *Cultura y Representaciones Sociales, 3*(6), 9–24.

Banco Mundial. (2007). *Informe sobre el desarrollo mundial 2008. Agricultura para el desarrollo*. Washington, DC: World Bank Group.

Bartra, A. (2006). *El capital en su laberinto. De la renta de la tierra a la renta de la vida*. Mexico: Editorial Itaca – Universidad Autónoma de la Ciudad de México.

Bartra, A. (2008). *El hombre de hierro. Los límites sociales y naturales del capital*. Mexico: Editorial Itaca.

O. F. Giraldo, *Political Ecology of Agriculture*,
https://doi.org/10.1007/978-3-030-11824-2

Bartra, A. (2013). *Hambre y Carnaval. Dos miradas a la crisis de la modernidad*. Mexico: Universidad Autónoma Metropolitana.

Bartra, A. (2016). Con los pies sobre la tierra. In A. Bartra, W. Porto-Gonçalves, & S. Betancourt (Eds.), *Se hace terruño al andar. Las luchas en defensa del territorio*. Mexico: ITACA.

Beigel, F. (2006). Vida, muerte y resurrección de las "teorías de la dependencia". In *Crítica y teoría en el pensamiento social latinoamericano*. Buenos Aires, Argentina: CLACSO.

Bejarano, F. (2003). Corporaciones, riesgos y prevención de daños de los plaguicidas. In F. Bejarano & B. Mata (Eds.), *Impactos del libre comercio, plaguicidas y transgénicos en la agricultura de américa latina*. Mexico: Red de acción sobre plaguicidas y alternativas en México.

Benjamin, W. (2008). *Tesis sobre la historia y otros fragmentos*. Mexico: ITACA-UACM.

Benjamin, W., & Millanes, J. M. (2004). *Sobre la fotografía*. Barcelona, Spain: Pre-textos.

Blaser, M. (2013). *Un relato de la globalización desde el Chaco*. Popayán, Colombia: Editorial Universidad del Cauca.

Boltvinik, J. (2009). Esbozo de una teoría de la pobreza y la sobrevivencia del campesinado. Polémica con Armando Bartra. *Mundo Siglo XXI*, (18), 27–41.

Borras, S. M., Jr., Hall, R., Scoones, I., White, B., & Wolford, W. (2011). Towards a better understanding of global land grabbing: An editorial introduction. *The Journal of Peasant Studies, 38*(2), 209–216.

Braudel, F. (1986). *La dinámica del capitalismo*. Mexico: Fondo de Cultura Económica.

Brenner, R. (1999). El desarrollo desigual y la larga fase descendente: Las economías capitalistas avanzadas desde el boom al estancamiento, 1950-1998. *Encuentro XXI, 14*, 199–370.

Brenner, R. (2009). Un análisis histórico-económico clásico de la actual crisis. *Revista Sin permiso*. Retrieved August 1, 2016, from http://www.sinpermiso.info/textos/index.php?id=2385

Broswimmer, F. J. (2005). *Ecocidio. Breve historia de la extinción en masa de las especies*. Pamplona, Mexico: Editorial Laeloti-Oceano.

Burkhart, E. (2012). De la economía a la amistad. In G. Esteva (Ed.), *Repensar el mundo con Iván Illich*. Guadalajara, Mexico: Taller editorial La Casa del Mago.

Capra, F. (1998). *La trama de la vida*. Barcelona, Spain: Editorial Anagrama.

Castro-Gómez, S. (2010). *Historia de la gubernamentalidad. Razón de Estado, liberalismo y neoliberalismo en Michel Foucault*. Bogotá, Colombia: Siglo del Hombre Editores-Pontificia Universidad Javeriana-Universidad Santo Tomás de Aquino.

Castro-Gómez, S. (2015). *Revoluciones sin sujeto. Slavoj Žižek y la crítica del historicismo posmoderno*. Mexico: Ediciones Akal.

Castro-Gómez, S., & Grosfoguel, R. (2007). Giro decolonial, teoría crítica y pensamiento heterárquico. In S. Castro-Gómez & R. Grosfoguel (Eds.), *El giro decolonial. Reflexiones para una diversidad epistémica más allá del capitalismo global*. Bogotá, Colombia: Siglo del Hombre Editores.

Composto, C., & Navarro, M. L. (2014). Claves de lectura para comprender el despojo y las luchas por los bienes comunes naturales en América Latina. In C. Composto & M. L. Navarro (Eds.), *Despojo capitalista, luchas en defensa de los bienes comunes naturales y alternativas emancipatorias para América Latina*. Mexico: Bajo Tierra Ediciones.

Corzo, A. M. (2011). *Del decreto al puente de los suspiros: impacto de la enfermedad de hansen (lepra) en las relaciones familiares de quienes la experimentaron. Agua de Dios Cundinamarca 1920–1960* (Master's thesis). Universidad Nacional de Colombia, Bogotá, Colombia.

Crowther, T. W., Glick, H. B., Covey, K. R., Bettigole, C., Maynard, D. S., Thomas, S. M., … Bradford, M. A. (2015). Mapping tree density at a global scale. *Nature, 525*(7568), 201–205.

Dagua, A., Aranda, M., & Vasco, L. G. (1998). *Guambianos. Hijos del Arcoiris y del agua*. Popayán, Colombia: CEREC.

Dazhong, W., & Pimentel, D. (1984). Energy flow through and organic agroecosystem in China. *Agriculture, Ecosystems & Environment, 11*(2), 145–160.

De Landa, M. (2011). *Mil años de historia no lineal*. Barcelona, Spain: Gedisa.

De Lisio, A. (2001). Del determinismo de la duración a la apertura del instante: Propuestas ante el pensamiento ambiental evolucionista. *Desenvolvimento e Meio Ambiente, 4*, 9–22.

Del Moral, J. M. (2004). *El nihilismo y el olvido del ser*. Texcoco, Mexico: Universidad Autónoma Chapingo.

Deleuze, G., & Guattari, F. (2004). *Mil mesetas. Capitalismo y esquizofrenia*. Valencia, Spain: Pre-Textos.

Descartes, R. (2008). *Discurso del método. Meditaciones metafísicas. Reglas para la dirección del espíritu. Principios de la filosofía*. Mexico: Editorial Porrúa.

Descola, P., & Pálsson, G. (Eds.). (2001). *Naturaleza y cultura. Perspectivas antropológicas*. Mexico: Siglo XXI Editores.

Díaz, S. (2011). El biopoder de la biotecnología o el biotecnopoder. Aportes para una bio(s) ética. *Ludus Vitalis, 19*(36), 193–211.

Duque, F. (1986). *Filosofía de la técnica de la naturaleza*. Madrid, Spain: Tecnos.

Dusch, G. (2017, March 02). Un acaparamiento de 10 millones. *La jornada*. Retrieved February 13, 2017, from http://www.jornada.unam.mx/2017/02/03/opinion/023a1pol

Eliade, M. (1981). *Lo sagrado y lo profano*. Barcelona: Guadarrama.

Escobar, A. (1996). Planificación. In W. Sach (Ed.), *Diccionario del desarrollo. Una guía del conocimiento como poder*. Lima, Peru: PRATEC.

Escobar, A. (1999). *El final del salvaje: naturaleza, cultura y política en la antropología contemporánea*. Bogotá, Colombia: CEREC.

Escobar, A. (2005) *Más allá del tercer mundo. Globalización y diferencia*. Bogotá, Colombia: Instituto Colombiano de Antropología e Historia.

Escobar, A. (2007). *La invención del Tercer Mundo. Construcción y deconstrucción del desarrollo*. Caracas, Venezuela: Fundación Editorial el Perro y la Rana.

Escobar, A. (2012). Cultura y diferencia: la ontología política del campo de Cultura y Desarrollo. *Wale'keru Revista de investigación en Cultura y Desarrollo, 2*, 8–29.

Escobar, A. (2013). En el trasfondo de nuestra cultura: la tradición racionalista y el problema del dualismo ontológico. *Tabula Rasa, 18*, 15–42.

Escobar, A. (2015). *Sentipensar con la tierra. Nuevas lecturas sobre desarrollo, territorio y diferencia*. Medellín, Colombia: Ediciones Unaula.

Escobar, A., & Osterweil, M. (2009). Movimientos sociales y la política de lo virtual. Estrategias deleuzianas. *Tabula Rasa, 10*, 123–161.

Espinoza, J., & Ortiz, P. (2014). La termodinámica de sistemas complejos alejados del equilibrio (TSCAE) como teoría y método en los estudios sociales. In A. Conde (Ed.), *Sobre Sistemas complejos, el pretendido fin*. Mexico City: Universidad Autónoma de Tlaxcala.

Estermann, J. (1998). *Filosofía andina. Estudio intercultural de la sabiduría autóctona andina*. Quito, Ecuador: Ediciones Abya-Yala.

Esteva, G. (2008). Agenda y sentido de los movimientos antisistémicos. *Polis Revista Latinoamericana*, (19), 8.

Esteva, G. (2009). *Más allá del desarrollo: la buena vida*. Quito, Ecuador: América Latina en Movimiento.

Esteva, G. (2011). Otra autonomía, otra democracia. In *Pensar las autonomías. Alternativas de emancipación al capital y al Estado*. Mexico: Sísifo Ediciones-Bajo Tierra.

Esteva, G. (2016). La casa común empieza con el techo común. In S. Street (Ed.), *Con ojos bien abiertos ante el despojo rehabilitemos lo común*. Guadalajara, Mexico: CIESAS.

Esteva, G. (2017, January 16). Aprender a gobernarnos. *La Jornada*. Retrieved January 16, 2017 from http://www.jornada.unam.mx/2017/01/16/opinion/016a1pol

Esteva, G., Prakash, M., & Stuchul, D. (2002). From a pedagogy for liberation to liberation from pedagogy. In V. Shiksha (Ed.), *Unfolding learning societies: Experiencing the possibilities*. Udaipur, Rajasthan, India: Shikshantar, The Peoples'Institute for Rethinking Education and Development..

ETC Group. (2009). *Who will feed us? Questions for the food and climate crisis. ETC Group Communiqué, 102*, 1.

FAO. (1954). *El estado mundial de la agricultura y la alimentación*. Rome, Italy: FAO.

FAO. (1996). *Enseñanzas de la revolución verde: hacia una nueva revolución verde. Documentos técnicos de referencia,* Rome. Retrieved April 19, 2015, from http://www.fao.org/docrep/003/w2612s/w2612s06.htm

FAO. (2008). *El estado mundial de la agricultura y la alimentación 2008*. Rome, Italy: FAO.

FAO. (2012a). *El estado mundial de la agricultura y la alimentación 2012*. Rome, Italy: FAO.

FAO. (2012b). *Dinámicas del Mercado de la Tierra en América Latina y el Caribe: Concentración y Extranjerización*. Rome, Italy: FAO.

FAO, & World Water Council (WWC). (2015). *Towards a water and food secure future. Critical Perspectives for Policy-makers*. Rome, Italy-Maseilla, France: FAO-WWC.

Fernandes, B. M. (2008). Entrando nos territórios do territoório. In E. T. Paulino & J. E. Fabrini (Eds.), *Campesinato e territórios em disputas*. Sao Paulo, Brazil: Expressão Popular.

Fernández, R., & González, L. (2014). *En la espiral de la energía. Tomo II: Colapso del capitalismo global y civilizatorio*. Madrid, Spain: Libros en acción.

Ford, A., & Nigh, R. (2016). *The Maya forest garden: Eight millennia of sustainable cultivation of the tropical woodlands*. London, England: Routledge.

Foucault, M. (2002). *Defender la sociedad*. Mexico: Fondo de Cultura Económica.

Foucault, M. (2006). *Seguridad, territorio, población*. Mexico: Siglo XXI Editores.

Foucault, M. (2009a). *El orden del discurso*. Mexico: Tusquets Editores.

Foucault, M. (2009b). *Vigilar y castigar*. Mexico: Siglo XXI Editores.

Foucault, M. (2014). *Los Anormales*. Buenos Aires, Argentina: Fondo de Cultura Económica.

Fromm, E. (1978). *¿Tener o ser?* Mexico: Fondo de Cultura Económica.

García-Canclini, N. (1990). *Culturas híbridas, estrategias para entrar y salir de la modernidad*. Mexico: Grijalbo.

Geertz, C. (1991). *La interpretación de las culturas*. Barcelona, Spain: Editorial Gedisa.

Giampetro, M., Cerretelli, G., & Pimentel, D. (1992). Energy analysis of agricultural ecosystem management: Human return and sustainability. *Agriculture, Ecosystems & Environment, 38*(3), 219–244.

Gibson-Graham, J. K. (1997). The end of capitalism (as we knew it): A feminist critique of political economy. *Capital & Class, 21*(2), 186–188.

Giraldo, O. F. (2012). El discurso moderno frente al "pachamamismo": La metáfora de la naturaleza como recurso y el de la Tierra como madre. *Polis Revista Latinoamericana, 11*(33), 219–233.

Giraldo, O. F. (2013). Hacia una ontología de la Agri-Cultura en perspectiva del pensamiento ambiental. *Polis Revista Latinoamericana, 34*, 95–115.

Giraldo, O. F. (2014). *Utopías en la era de la supervivencia. Una interpretación del Buen Vivir*. Mexico: Editorial Itaca-Universidad Autónoma Chapingo.

Giraldo, O. F. (2015). Agroextractivismo y acaparamiento de tierras en América Latina: Una lectura desde la ecología política. *Revista Mexicana de Sociología, 77*(4), 637–662.

Giraldo, O. F. (2016a). Geopoéticas de la agri-cultura y el agroextractivismo industrial: La pregunta por el habitar. *Geograficidade, 5*(2), 76–88.

Giraldo, O. F. (2016b). Convivialidad y agroecología. In S. Street (Ed.), *Con ojos bien abiertos ante el despojo rehabilitemos lo común*. Guadalajara, Mexico: CIESAS.

Giraldo, O. F., & Rosset, P. M. (2017). Agroecology as a territory in dispute: Between institutionality and social movements. *Journal of Peasant Studies, 45*(3), 545–564.

Giraldo, O. F., & Vanegas, A. Agroecology and resilience of socio-ecological systems: A thermodynamic approach. Unpublished.

Gliessman, S. (1998). *Agroecology: Ecological processes in sustainable agriculture*. Ann Arbor, MI: Sleeping Bear-Ann Arbor Press.

Gliessman, S., Rosado-May, F., Guadarrama-Zugasti, C., Jedlicka, J., Cohn, A., Mendez, V., … Jaffe, R. (2007). Agroecología: promoviendo una transición hacia la sostenibilidad. *Ecosistemas: Revista Cietifica y Tecnica de Ecologia y Medio Ambiente, 16*(1), 13–23.

Global Witness. (2015). *¿Cuantos más?* London: Global Witness.

Gonnet, C. O. (2015). Mecanismos de difusión de los Programas de Transferencia Condicionada en América Latina. El caso chileno. *Íconos. Revista de Ciencias Sociales*, (53), 31–48.

González-Jacome, A., del Amo, S., & Gurri, F. (2007). *Los nuevos caminos de la agricultura: procesos de conversión y perspectivas. Plaza y Valdés*. Mexico City: Iberoamerican University.

Grain. (2009, October, 9–16). Earth matters: Tackling the climate crisis from the ground up. *Seedling*.

Grain. (2013). Commentary IV: Food, climate change and healthy soils: The forgotten link. In *Trade and environment review 2013. Wake up before it is too late*. Genova, Italy: UNCTAD.

Grain. (2016). El acaparamiento global de tierras en el 2016: sigue creciendo y sigue siendo malo. In *A Contrapelo*. Barcelona, Spain: Grain.

Grain. (2017). *Cultivando desastres. Las principales compañías del mundo van tras la cosecha*. Barcelona, Spain: Grain.

Grifon, D. (2012). *Reflexiones Sobre los niveles de la Agroecología.*

Gronemeyer, M. (1996). Ayuda. In W. Sach (Ed.), *Diccionario del desarrollo. Una guía del conocimiento como poder*. Lima, Peru: PRATEC.

Grosfoguel, R. (2016). Del "extractivismo económico" al "extractivismo epistémico" y al "extractivismo ontológico": una forma destructiva de conocer, ser y estar en el mundo. *Tabula Rasa, 24*, 123–143.

Gudynas, E. (2009). Diez tesis urgentes sobre el nuevo extractivismo. Contextos y demandas bajo el progresismo sudamericano actual. In *Extractivismo, política y sociedad*. Quito, Ecuador: CAAP-CLAES.

Gudynas, E. (2011). Buen Vivir: Germinando alternativas al desarrollo. *América Latina en movimiento, 462*, 1–20.

Haesbaert, R. (2011). *El mito de la desterritorialización. Del fin de los territorios a la multiterritorialidad*. Mexico: Siglo XXI Editores.

Harvey, D. (2003). *Espacios de esperanza*. Madrid, Spain: Ediciones Akal.

Harvey, D. (2007). *El nuevo imperialismo*. Madrid, Spain: Ediciones Akal.

Heidegger, M. (1994a). Aletheia (Heráclito-Fragmento 16). In *Conferencias y Artículos*. Barcelona, Spain: Ediciones del Serbal.

Heidegger, M. (1994b). La pregunta por la técnica. In *Conferencias y Artículos*. Barcelona, Spain: Ediciones del Serbal.

Heidegger, M. (1994c). Superación de la metafísica. In *Conferencias y Artículos*. Barcelona, Spain: Ediciones del Serbal.

Heidegger, M. (1994d). Construir, habitar, pensar. In *Conferencias y Artículos*. Barcelona, Spain: Ediciones del Serbal.

Heidegger, M. (1996). La época de la imagen del mundo. In *Caminos del bosque*. Madrid, Spain: Alianza.

Hernández-Castillo, R., & Nigh, R. (1998). Global processes and local identity among mayan coffee growers in Chiapas, Mexico. *American Anthropologist, 100*(1), 136–147.

Hinkelammert, F. (2002). *El retorno del sujeto reprimido*. Bogotá, Colombia: Universidad Nacional de Colombia.

Holleman, H. (2016). De-naturalizing ecological disaster: Colonialism, racism and the global Dust Bowl of the 1930s. *The Journal of Peasant Studies, 44*(1), 234–260.

Holloway, J. (2005). *Cambiar el mundo sin tomar el poder: el significado de la revolución hoy*. Valencia, Spain: Vadell Hermanos Editores.

Holloway, J. (2011). *Agrietar el capitalismo: el hacer contra el trabajo*. Madrid, Spain: El Viejo Topo.

Holt-Gimenez, E. (2008). *Campesino a campesino. Voces de Latinoamérica. Movimiento Campesino a Campesino para la agricultura sustentable*. Managua, Nicaragua: SIMAS.

Holt-Giménez, E., & Patel, R. (2009). *Rebeliones alimentarias. Crisis y hambre de justicia*. Oakland, CA: Food First.

Horkheimmer, M. (2002). *Crítica de la razón instrumental*. Madrid, Spain: Editorial Trotta.

Huxley, A. (2004). *Un mundo feliz*. Bogotá, Colombia: Casa Editorial El Tiempo.

Illich, I. (1996). Necesidades. In W. Sach (Ed.), *Diccionario del desarrollo. Una guía del conocimiento como poder*. Lima, Peru: PRATEC.

Illich, I. (2006a). Alternativas. In I. Illich (Ed.), *Obras reunidas I*. Mexico: Fondo de Cultura Económica.

Illich, I. (2006b). La convivencialidad. In I. Illich (Ed.), *Obras reunidas I*. Mexico: Fondo de Cultura Económica.

Illich, I. (2013). *La sociedad desescolarizada*. Buenos Aires, Argentina: Ediciones Godot.

Ingold, T. (1990). Society, nature and the concept of technology. *Archaeological Review from Cambridge, 9*(1), 5–17.

Ingold, T. (2000). *The perception of the environment. Essays on livelihood, dwelling and skill*. London, England/New York, NY: Routledge.

Janke, W. (1988). *Postontología*. Bogotá D.C., Colombia: Pontificia Universidad Javeriana-Organización de los Estados Iberoamericanos para la Educación la Ciencia y la Cultura.

Jullien, F. (2005). *Del "tiempo". Elementos de una filosofía del vivir*. Madrid, Spain: Arena Libros.

Kangmennaang, J., Bezner-Kerr, R., Lupafya, E., Dakishoni, L., Katundu, M., & Luginaah, I. (2017). Impact of a participatory agroecological development Project on household wealth and food security in Malawi. *Food Security, 9*(3), 561–576.

Khadse, A., Rosset, P. M., Morales, H., & Ferguson, B. G. (2017). Taking agroecology to scale: The zero budget natural farming peasant movement in Karnataka, India. *The Journal of Peasant Studies, 45*(1), 192–219.

Kirwan, A. D. (2008). Quantum and ecosystem entropies. *Entropy, 10*(2), 58–70.

Koohafkan, P., & Altieri, M. (2010). *Sistemas Importantes del patrimonio Agrícola Mundial. Un Legado para el Futuro*. Rome, Italy: FAO.

Kotschi, J. (2013). *A soiled reputation: Adverse impacts of mineral fertilizers in tropical agriculture*. Berlin, Germany: WWF International-Heinrich Böll Stiftung.

Lander, E. (Ed.). (2000). *La colonialidad del saber: eurocentrismo y ciencias sociales. Perspectivas latinoamericanas*. Buenos Aires, Argentina: CLACSO.

Landmatrix. (2017). Retrieved February 10, 2017, from http://www.landmatrix.org/en/

Latouche, S. (2007). *La otra áfrica. Autogestión y apaño frente al mercado global*. Barcelona, Spain: Oozebap.

Lefevbre, H. (2013). *La producción del espacio*. Madrid, Spain: Capitán Swing.

Leff, E. (2004). *Racionalidad ambiental. La reapropiación social de la naturaleza*. Mexico: Siglo XXI Editores.

Leff, E. (2008). Decrecimiento o desconstrucción de la economía: Hacia un mundo sustentable. *Polis Revista Latinoamericana, 7*(21), 81–90.

Leff, E. (2014). *La apuesta por la vida. Imaginación sociológica e imaginarios sociales en los territorios ambientales del sur*. Mexico: Siglo XXI Editores.

Lefort, C. (2004). *La incertidumbre democrática: ensayos sobre lo político*. Barcelona, Spain: Anthropos Editorial.

León, E. (2011). *El monstro en el otro. Sensibilidad y coexistencia humana*. Madrid, Spain: Sequitur.

Lin, B. B., Jahi Chappell, M., Vandermeer, J., Smith, G., Quintero, G., Bezner-Kerr, R., … Perfecto, I. (2011). Effects of industrial agriculture on climate change and the mitigation potential of small-scale agro-ecological farms. *Animal Science Reviews, 6*(20), 1–18.

Loh, J. (2000). *Living planet report 2000*. Gland, Switzerland: WWF International.

Luxemburgo, R. (1967). *La acumulación del capital*. Mexico D.F., Mexico: Grijalbo.

LVC (La Vía Campesina). (2015a). *Declaración Foro Internacional de Agroecología*. Retrieved June 5, 2016, from http://viacampesina.org/es/index.php/temas-principales-mainmenu-27/agricultura-campesina-sostenible-mainmenu-42/2354-declaracion-del-foro-internacional-de-agroecologia

LVC (La Vía Campesina). (2015b). *Peasant agroecology for food sovereignty and mother earth, experiences of La Vía Campesina*. Notebook no. 7. Zimbabwe.

LVC (La Vía Campesina). (2017). *Toolkit: Peasant agroecology schools and the peasant-to-peasant method of horizontal learning*.

Machado, H. A. (2014). Territorios y cuerpos en disputa. Extractivismo minero y ecología política de las emociones. *Revista sociológica de pensamiento crítico, 8*(1), 56–71.

Machín, B., Roque, A. M., Ávila, D. R., & Rosset, P. M. (2010). *Revolución agroecológica. El movimiento campesino a campesino de la ANAP en Cuba*. Bogotá D.C., Colombia: Asociación Nacional de Agricultores Pequeños-La Vía Campesina.

Madelin, P. (2016). *Après le capitalisme. Essai d'écologie politique*. Montreal, Canada: Écosociété.

Mandel, E. (1972). *El capitalismo tardío*. Mexico: Ediciones Era.

Marañon, B. (2013). *La economía solidaria en México: Una primera aproximación. Hacia una alternativa societal basada en la reciprocidad*. Mexico: Universidad Nacional Autónoma de México.

Marcuse, H. (1986). *El hombre unidimensional*. Mexico: Editorial Joaquín Mortiz.

Mardones, F. (2016). El espacio para una nueva forma de pensar nuestra economía. *Polis Revista Latinoamericana, 15*(45), 121–139.

Mariaca, R. (2010). La agrobiodiversidad: ¿sabemos cuántas plantas se cultivan y cuántos animales se crían en el sureste de México? *Ecofronteras, 40*, 10–13.

Marx, K. (1946). *El capital. Crítica de la economía política. Tomo II*. Bogotá D.C., Colombia: Fondo de Cultura Económica.

Marx, K. (1974). Proyecto de respuesta a la carta de V. I. Zasulich. In C. Marx & F. Engels (Eds.), *Obras Escogidas en tres tomos. Tomo I*. Moscow, Russia: Editorial Progreso.

Marx, K., & Engels, F. (2001). *Manifiesto del partido comunista*. Bogotá, Colombia: Panamericana Editorial.

Maturana, H. (2007). *Emociones y lenguaje en educación y política*. Santiago de Chile, Chile: Comunicaciones Noreste.

Maturana, H., & Dávila, X. (2013). La gran oportunidad: fin de la psiquis del liderazgo en el surgimiento de la psiquis de la gerencia co-inspirativa. *Revista chilena de administración pública, 10*, 101–124.

Maturana, H., & Varela, F. (2003). *El árbol del conocimiento. Las bases biológicas del conocimiento humano*. Buenos Aires, Argentina: Lumen.

Mayer, A. L., Donovan, R. P., & Pawlowski, C. W. (2014). Information and entropy theory for the sustainability of coupled human and natural systems. *Ecology and Society, 19*(3), 11.

McMichael, P (2015) *Regímenes alimentarios y cuestiones agrarias*. Mexico: Miguel Ángel Porrua.

Mier y Terán Giménez Cacho, M., Giraldo, O. F., Aldasoro, M., Morales, H., Ferguson, B. G., Rosset, P., & Campos, C. (2018). Bringing agroecology to scale: Key drivers and emblematic cases. *Agroecology and Sustainable Food Systems, 42*(6), 637–665.

Millennium Ecosystem Assessment (MEA). (2005). *Ecosystems and human well-being: Synthesis*. Washington, DC: Island Press.

Moore, J. W. (2017). The Capitalocene, Part I: On the nature and origins of our ecological crisis. *The Journal of Peasant Studies, 44*(3), 594–630.

Navarro, M. L. (2015). *Luchas por lo común. Antagonismo social contra el despojo capitalista de los bienes naturales en México*. Mexico: ICSyH BUAP-Bajo tierra ediciones.

Ng, F., & Ataman, M. A. (2008). *Who are the net food importing countries?* Washington, DC: Banco Mundial.

Nhat Hanh, T. (1998). *Las claves del zen*. Madrid, Spain: Neo Person.

Nietszche, F. (1999). *La gaya ciencia*. Madrid, Spain: Alba Libros.

Nietszche, F. (2000). Así Hablaba Zaratustra. In *Obras Inmortales. Tomo 2*. Barcelona, Spain: Edicomunicación.

Nisbet, R. (1981). *Historia de la idea de progreso*. Barcelona, Spain: Editorial Gedisa.

Noguera, A. P. (2004). *El reencantamiento del mundo*. Mexico-Bogotá DC, Colombia: PNUMA-Universidad Nacional de Colombia.

Noguera, A. P. (2012). *Cuerpo-Tierra. El Enigma, El Habitar, La vida. Potencias de un Pensamiento Ambiental en clave del Reencantamiento del Mundo*. Madrid, Spain: Editorial Académica Española.

Noguera, A. P., & Bernal, D. A. (2013). Tensiones entre el mundo tecnológico y el mundo de la vida. *Logos, 23*, 21–37.

Noguera, A. P., & Giraldo, O. F. (2017). ¿Para qué poetas en tiempos de extractivismo ambiental? In H. Alimonda, C. Toro, & F. Martín (Eds.), *Ecología Política Latinoamericana. Pensamiento crítico, diferencia latinoamericana y rearticulación epistémica*. Buenos Aires, Argentina: CLACSO.

O'Connor, J. (2001). *Causas naturales. Ensayos de marxismo ecológico*. Mexico: Siglo XXI Editores.

Osorio, C. (2015). Mecanismos de difusión de los Programas de Transferencia Condicionada en América Latina: el caso chileno. *Iconos, 53*, 31–48.

Oxfam. (2011). *Tierra y Poder. El creciente escándalo en torno a una nueva oleada de inversiones en tierra*. Oxford, England: Oxfam.

Oxfam. (2012). *Nuestra tierra, nuestras vidas*. Oxford, England: Oxfam.

Pardo, J. L. (1991). *Sobre los espacios Pintar, escribir, pensar*. Barcelona, Spain: Ediciones del Serbal.

Perfecto, I., & Vandermeer, J. (2010). The agroecological matrix as alternative to the land-sparing/agriculture intensification model. *Proceedings of the National Academy of Sciences, 107*(13), 5786–5791.

Pineda, J. (2014a). *"El pensamiento ambiental" Cedum, Universidad de Manizales*. Retrieved April 19, 2015, from https://www.youtube.com/watch?v=_WM8SePC-Vs

Pineda, J. (2014b). *Geopoética de la guerra* (PhD thesis in social sciences, childhood and youth). CINDE-Universidad de Manizales, Manizales, Colombia.

Pineda, J. (2016). Paisajes del desarrollo: desilusión, disolución, devastación y desolación. In A. P. Noguera (Ed.), *Voces del pensamiento ambiental. Tensiones críticas entre desarrollo y Abya Yala*. Bogotá, Colombia: Universidad Nacional de Colombia.

Pingali, P. L., Hossain, M., & Gerpacio, R. V. (1997). *Asian rice bowls: The returning crisis?* Manila, Philippines: International Rice Research Institute.

Platón. (1958). *La República o el Estado*. Mexico: Austral.

Ploeg, J. D. (2010). *Nuevos campesinos. Campesinos e Imperios alimentarios*. Barcelona, Spain: Editorial Icaria.

Polanyi, K. (1975). *La gran transformación*. Mexico: Juan Pablos Editor.

Porto-Gonçalves, C. A., & Leff, E. (2015). Political ecology in Latin America: The social re-appropriation of nature, the reinvention of territories and the construction of an environmental rationality. *Desenvolvimiento e Meio Ambiente, 35*, 65–88.

Porto-Gonçalves, C. W. (2002). Latifundios genéticos y existencia indígena. *Revista Chiapas, 14*, 7–30.

Porto-Gonçalves, C. W. (2006). *El desafío ambiental*. Mexico: PNUMA.

Prigogine, I., & Stengers, I. (1984). *Self-organization in non-equilibrium systems*. New York, NY: Wiley.

Quijano, A. (2000). Colonialidad del poder, eurocentrismo y América Latina. In E. Lander (Ed.), *La colonialidad del saber: eurocentrismo y ciencias sociales. Perspectivas latinoamericanas*. Buenos Aires, Argentina: CLACSO.

Rahnema, M. (1996). Participación. In W. Sach (Ed.), *Diccionario del desarrollo. Una guía del conocimiento como poder*. Lima, Peru: PRATEC.

Randall, J. (1952). *La formación del pensamiento moderno*. Buenos Aires, Argentina: Editorial Nova.

Ray, D. K., Ramankutty, N., Mueller, N. D., West, P. C., & Foley, J. A. (2012). Recent patterns of crop yield growth and stagnation. *Nature Communications, 3*, 1293.

Ricoeur, P. (1980). *La metáfora viva*. Madrid, Spain: Ediciones Cristiandad.

Robert, J. (2012). Un filósofo y pensador radical en Cuernavaca. In G. Esteva (Ed.), *Repensar el mundo con Iván Illich*. Guadalajara, Mexico: Taller editorial La Casa del Mago.

Robert, J., & Rahnema, M. (2015). *La potencia de los pobres*. San Cristóbal de las Casas, Mexico: CIDECI-UNITIERRA.

Robles, H., & Concheiro, L. (2014). Vigencia de la reforma agraria en América Latina. In F. Hidalgo, F. Houtart, & P. Lizárraga (Eds.), *Agriculturas campesinas en Latinoamérica. Propuestas y desafíos*. Quito, Ecuador: Instituto de Altos Estudios Nacionales, IAEN.

Rockström, J., Steffen, W., Noone, K., Persson, Å., Chapin, F. S., III, Lambin, E., … Foley, J. (2009). Planetary boundaries: Exploring the safe operating space for humanity. *Ecology and Society, 14*(2), 32.

Rorry, R. (2009). *La filosofía de la Naturaleza*. Madrid, Spain: Cátedra.

Rosset, P., & Martínez-Torres, M. E. (2012). Rural social movements and agroecology: Context, theory, and process. *Ecology and society, 17*(3), 17.

Rosset, P. M. (2016). La reforma agraria, la tierra y el territorio: Evolución del pensamiento de La Vía Campesina. *Mundo agrario, 17*(35), e021.

Roudinesco, E. (2000). *¿Por qué el psicoanálisis?* Madrid, Spain: Paidós.

Rulfo, J. (2001). *Pedro Páramo*. Barcelona, Spain: Anagrama.

Running, S. W. (2012). A measurable planetary boundary for the biosphere. *Science, 337*, 1458.

Santos, B. S. (2009). *Una epistemología del sur: la reinvención del conocimiento y la emancipación social*. Mexico: Siglo XXI Editores-CLACSO.

Sbert, J. M. (1996). Progreso. In W. Sach (Ed.), *Diccionario del desarrollo. Una guía del conocimiento como poder*. Lima, Peru: PRATEC.

Schneider, E., & Sagan, D. (2008). *La termodinámica de la vida*. Barcelona, Spain: Tusquets.

Schrödinger, E. (2005). *¿Qué es la vida?* Salamanca, Spain: Universidad de Salamanca.

Schüssler, I. (1998). *La tierra y lo sagrado*. Texcoco, Mexico: Universidad Autónoma Chapingo.

Scott, J. C. (2000). *Los dominados y el arte de la resistencia. Discursos ocultos*. Mexico: Ediciones Era.

Serres. (1990). *El contrato natural*. Valencia, Spain: Pre-Textos.

Shiva, V. (2007). *Los monocultivos de la mente*. Monterrey, Mexico: Editorial Fineo.

Soto-Torres, G. (2015). Aforismos sobre el habitar humano-naturaleza. *Geograficidade, 5*, 208–213.

Spinoza, B. (2011). *Ética*. Madrid, Spain: Alianza Editorial.

Sutcliffe, B. (1995). Desarrollo frente a ecología. *Ecología política, 9*, 27–47.

Tainter, J. (1988). *The collapse of complex societies*. Cambridge, England: Cambridge University Press.

Toledo, V. (2008). Metabolismos rurales: hacia una teoría económico-ecológica de la apropiación de la naturaleza. *Revista Iberoamericana de Economía Ecológica, 7*, 1–26.

Toro, I. *Empatía y posdesarrollo*. Tesis de Maestría, CESMECA-UNICACH, San Cristóbal de las Casas. Unpublished.

Torres, G. (2006). *Poscivilización: Guerra y ruralidad*. Mexico: Plaza y Valdés.

Toussaint, O., & Schneider, E. D. (1998). The thermodynamics and evolution of complexity in biological systems. *Comparative Biochemistry and Physiology Part A: Molecular & Integrative Physiology, 120*(1), 3–9.

Tyrtania, L. (2008). La indeterminación entrópica: Notas sobre disipación de energía, evolución y complejidad. *Desacatos, 28*, 41–68.

Ulloa, A. (2004). *La construcción del nativo ecológico. Complejidades, paradojas y dilemas de la relación entre los movimientos indígenas y el ambientalismo en Colombia*. Bogotá, Colombia: Instituto Colombiano de Antropología e Historia-COLCIENCIAS.

Varela, F. (1998). *El sueño, los sueños y la muerte. Exploración de la conciencia con S.S. El Dalai Lama*. Barcelona, Spain: José J. de Olañeta.

Varela, F. (2000). *El fenómeno de la vida*. Santiago de Chile, Chile: Dolmen Ediciones.

Varela, F. (2004). *Monte grande*, Documental de Reichle, F., Zürich, t & c Zürich.

Varela, F., Thompson, E., & Rosch, E. (1997). *De cuerpo presente. Las ciencias cognitivas y la experiencia humana*. Barcelona, Spain: Editorial Gedisa.

Vásquez, J. I., & Rivas, A. (2006). *De campesino a campesino*. Managua, Nicaragua: Unión Nacional de Agricultores y Ganaderos.

Vattimo, G. (1998). *El fin de la modernidad. Nihilismo y hermenéutica en la cultura posmoderna*. Barcelona, Spain: Gedisa.

Vergara-Camus, L., & Kay, C. (2017). Agribusiness, peasants, left-wing governments, and the state in Latin America: An overview and theoretical reflections. *Journal of Agrarian Change, 17*, 239–257.

Wallerstein, I. (2001). *Después del liberalismo*. Mexico: Siglo XXI Editores.

Wallerstein, I. (2010). *El capitalismo histórico*. Mexico: Siglo XXI Editores.

Waters, C. N., Zalasiewicz, J., Summerhayes, C., Barnosky, A. D., Poirier, C., Gałuszka, A., & Jeandel, C. (2016). The Anthropocene is functionally and stratigraphically distinct from the Holocene. *Science, 351*(6269), aad2622.

Wittgenstein, L. (1988). *Investigaciones filosóficas*. Barcelona, Spain: Editorial Crítica.

Wolf, E. (2001). *Figurar el poder: Ideologías de dominación y crisis*. Mexico: CIESAS.

World Bank (2007). World Development Report. Agriculture for development. Washington D.C.

World Economic Forum. (2017). New Vision for Agriculture.

World Wildlife Fund for Nature (WWF) (2009). *Palma de Aceite. Documento de Posición de WWF Colombia*. Bogotá D.C., Colombia: WWF Colombia.

Worster, D. (1979). *Dust bowl: The southern plains in the 1930s*. New York, NY: Oxford University Press.

Worster, D. (2008). *Transformaciones de la tierra*. Montevideo, Uruguay: Coscoraba Ediciones.

WWF. (2014). *El crecimiento de la soja: Impactos y soluciones*. Gland, Switzerland: WWF International.

Ziga, F. (2013). *Ritualidad y temporalidad en horizontes culturales diferenciados: la costa chica de Oaxaca* (PhD thesis in agricultural sciences). Universidad Autónoma Chapingo, Texcoco, Mexico.

Žižek, S. (1992). *El sublime objeto de la ideología*. Mexico: Siglo XXI Editores.

Index

A
Accumulation by dispossession, 38, 39, 58, 61, 74, 78, 119
Adaptations, 98, 99
Affections, 61
Agribusiness, 5, 7, 8, 17, 29
  agricultural variety, 58
  biopolitics, 66
  dispossession, 58
  investor, 57
  on land, agricultural activity, 57
  policies, 57
  and territorial rent, 57
Agricultural activity, 57
Agricultural capitalism, 8, 31, 88
Agricultural machinery, 7
Agricultural over-accumulation
  agricultural oversupply, 24
  colonialism, 24
  development, 25
  "Development of Underdevelopment" project, 24
  European demand, 24
  fertilizer dissemination program, 27
  food regime, 23
  free trade agreements, 28
  global food regime, 28
  green revolution, 27
  poverty, 24, 25
  poverty-representation regime, 25, 26
  programs, 25
  WFP, 26
  "Wheat for Oil", 27
Agricultural proletariats, 73
Agricultural revolution, 63
Agriculture
  co-evolution, 99
  co-evolutionary process, 100
  "create linkages", 63
  culture and environment, 101
  and culture role, 100
  ecological science, 101
  economic rationality, 21
  and natural drift, 98, 99
  natural-drift thinking, 99
  ontological transformation, 63, 66
  revolution, 100
Agro-biotechnology, 109
Agro-capitalism
  civilizing progress, 12
  epistemic colonization, 14
  extractivism, 13
  flexibility, 31
  future-oriented temporal rationality views, 14
  human destiny, 11
  metaphysical rationality, journey, 12
  modernity, 12
  modernization, 13
  temporality, 11
  territorial growth, 35
  time, 11
Agroecological complexity, 109
Agroecological principles, 105
Agroecological technique, 113
Agroecology, 58, 75, 76
  in agenda of international organizations, 93
  agricultural capitalism, 88
  agroecological movements, 92
  and autopoiesis, 104, 105

O. F. Giraldo, *Political Ecology of Agriculture*,
https://doi.org/10.1007/978-3-030-11824-2

Agroecology (*cont.*)
  capitalism, 91
  communal tradition, 87
  fruits, peasant-to-peasant, 87
  heterarchical trials, 92
  natural drift, 102
  second law of thermodynamics (*see* The second law of thermodynamics)
  social processes (*see* Social processes, agroecology)
Agro-extractive activity, 10, 30
Agro-extractive model, 17
Agro-extractivism
  and agroecological conditions, 99
  biotechnology, 4
  civilizational crisis, 97
  critique, 1
  deep-rooted certainty, 1
  economic rationality (*see* Economic rationality)
  entrepreneurial education, 51
  interventions, 3
  metaphysics and economic rationality, 15
  rationality, 17
  rural development, 14
  'Third World' countries, 13, 14
  "Western", 1
Agro-forestry management, 111
Agro-industrial sector, 32
Agro-industrial system, 29
Agro-industry, 29
Agro-technology, 114
Alliance for Progress, 25
Anthropic ecocide, 62
Anthropocene, 78
Autopoiesis, 104, 105

B
Biocultural co-evolution, 121
Biofuels, 49
Biopolitics, 66, 92, 129
Biopower, 9
Biotechnological science, 9
Bio-trade, 31, 58
Bird, 100

C
Capital accumulation, 64
Capitalism, 31
  bio-trade, 58
  crisis of, 36
  cultural elements, 43
  development, 37
  discourse, 68
  geographical expansion, 37
  globalized, 37
  over-accumulation, 37, 38
  "peaceful" strategy, 40
  small producers' lands, 42
Capitalist modernity, 43, 75, 94, 123
Capitalist production, 29
Capitalist system, 48, 91
Capitalist-industrial civilization, 118
Center-periphery dialectic, 37
Civilization, 121
Climate-smart agriculture, 31
Co-evolution, 99–101
"Coloniality of power", 1
Colonization
  epistemic, 45
  ontological, 43
Commodities, 4, 8, 10
Complementarity, 2
Contract farming, 49, 52
Conventional oil, 118
Creation of linkages, 62, 63
Creativity, 112–114
Crisis of civilization, 130

D
Darkness, 2, 3
Death policies, 125
Degrowth, 86
De-humanization, 65
Democratic society, 68
Desires, 68–70
Deterritorialization, 47, 52, 56, 61, 69
"Development of Underdevelopment" project, 24
Differential rent, 57
Disseminated agroecological practices, 83

E
Earth of agroextractivism's linear process, 104
Economic and social heterogeneity, 50
Economic liberalism, 19
Economic rationality, 42, 50, 64
  and agriculture, 21
  capitalism, 21
  capitalist development, 21
  colonial process, 22
  economic liberalism, 19
  economic science, 19, 20
  economic system, 18

economized geographies, 28
free-market system, 19
imperialist agrarian expansion, 23
market economy, 18
metaphysical thinking, 18
over-accumulation, 23, 26 (*see also* Agricultural over-accumulation)
self-regulated economy, 19
social system, 21
traditional agricultural systems, 22
Economic science, 19, 20, 64
Economy, 19, 20, 28
Ecosystemic simplification, 76, 108
End of the fossil fuel era, 119
English capitalism
in Latin America, 22
Entropy, 106, 107, 109, 111
Enunciation fields, 53
Environmental debacle, 31
Environmental ethics, 20
Environmental history of agriculture, 17, 18
Environmental service, 54
Equilibrium, 107
Esthetics, agro-industrial
corporality, 71
definition, 70
development undertaken, 74
ecosystemic transformations, 71
empirical evidence, 72
exercise of power, 73
geo-graphies, 73
green revolution, 71
modernity, 72
mutual affectation, 71
perception changes, 72
planned landscapes, 71
worth mentioning, 73
Exercise of power, 47, 52, 73, 94
Expropriation practices, 70
Extractive agribusiness, 67, 97, 105, 108, 128
Extractivism, 78, 127
agroextractivism's interventions, 3
output, 4
productivity, 4
Extractivist policies, 1

F

Farmers, global South, 44
Fertilizer dissemination program, 27
Financialization, 38
First contradiction of capital, 36
Food regime, 23, 28
Food sovereignty, 32, 78
Fossil-fuel-based agriculture, 117
Foucault, M., 47, 48, 56, 65, 66
Free-market system, 19
Futures and agroecology, 120, 121, 123, 124

G

Genes, 9
Global Campaign against Hunger, 27
Global capitalism, 43
Global warming, 28, 30, 108
Governmentalization, 35
Grassroots organizations, 78, 92, 132
Great development project, 86
Green deserts, 3
Green plants, 106
Green revolution, 7, 9, 27, 32

H

Hegemonic economic science, 21
Heterarchical trials, 92
Hoarding, 40
Homo economicus, 20, 51
Human creativity, 103, 105
Human destiny, 11
Human habitation, 62, 94, 110
Human resources, 117

I

"I-ist" habits, 68
In situ deterritorialization, 46, 47
Industrial agribusiness, 14, 49
Industrial agriculture, 117, 119
Industrial agroextractivism, 109, 130
Industrial food system, 76
Industrial society, 119
Intellectual property rights, 39
Irrigation, 7

L

Labor force, 22, 23, 48
Laboratory-invented eco-technologies, 114
Land dispossession, 51
accumulated land, 39
soil, 40
Land-grabbing, 46, 49, 70
Land sharing, 126
Land sparing, 125
Landless Rural Workers Movement (MST) in Brazil, 77
Language, 53, 54

Large-scale agroecological reconversion, 128
Large-scale corporate investment, 49
Lazarettos, 47
Leprosy control, 47, 48
Liberal economy, 28
Liberation theology, 79, 80, 93
Life, 104, 105
Linkages, 62–65, 67

M
"Market-based" program, 52
Mayan civilization, 121
Metaphysical thinking
  agro-biotechnological activity, 9
  agro-extractive business, landscapes, 10
  artificial nature, 10
  Christianity, 6
  and green revolution, 7, 9
  "metaphysics", 5
  in modernity, 6, 7
  nature, 8
  neoplatonism, 6
  Plato's philosophy, 5, 6
  scientific-technical methods, 10
  Western culture, 10
Metaphysical Western culture, 8
Metaphysics, 2, 5
Millennial Agri-Culture, 44
Millennium Ecosystem Assessment (MEA) report, 29
Milpa system, 122
Modern rationality, 17, 18
Modernity, 12, 72
Modernization, 13
Mutual affectation, 71

N
Natural agricultural techniques, 84
Natural drift, 98, 99, 102
Nature, 8, 104
Nature/culture separation, 55
Negentropy, 107, 108, 110
Neoliberalism, 78
Neoplatonism, 6
Nitrogen fertilizers, 30
Non-equilibrium, 107

O
Oil-based agro-capitalism, 108
Ontological differences, 43
Ontological transformation, 127–129, 132
Original accumulation, 38, 40–41, 48
Over-accumulation, 37

P
Participatory development, 69
Peasant economies, 45
Peasant knowledge, 44, 45
Peasant-to-peasant
  fruits, 87
  grassroots network, 88
  hierarchical and network components, 89
  horizontal exchange of wisdom, 85
  liberation theology, 80
  mobilizing methodologies, 132
  Peasant-to-Peasant Movement, 81
  post-development, 89
  traditional wisdom, 87
  Zero Budget Spiritual Farming effort in India, 84
Pesticides, 30, 31
Photosynthesis, 106, 107, 111
Plague-control model, 48
Plantations, 49, 50
Plato's philosophy, 5, 6
Policies, 70
Political administration of affections
  agricultural revolution, 63
  anthropic ecocide, 62
  biopolitics, 66
  capital accumulation, 64
  "create linkages", 63
  creation, agriculture, 62
  economic rationality, 64
  emotional structure, 65
  ontological transformation, 63
  relational ontologies, 64
  social life, 63
  territorial control, 66
Political ecology, 130
Polycultures, 44
Post-development
  agriculture, 98, 129
  autonomy, 89
  economy, 89
  good living, 85
  Living Well project, 131
  natural drift, theory of, 99
  non-agricultural movements, 94
  peasant-to-peasant, 80, 85, 132
  polycentric network and internetwork designs, 88
  and post-extractivism, 103, 133
  principles, 86

rebuilding techniques, 114
social movements, 86
unpredictability, 89
Poverty, 24, 25, 67
Precision agriculture, 31
Privatization, 38, 39
Producing desire
affectation of desire, 70
affective deterritorialization, 69
capitalism's discourse, 68
democratic society, 68
development projects, 69
green revolution, 67
"I-ist" habits, 68
multinational mining and petroleum corporations, 69
policies, 70
poverty, 67
psychoanalysis, 66
rural poverty, 67
Productivity, 4
Psychoanalysis, 66, 94

R
Rationality, 64
Relational ontologies, 64, 131
Rural benefactors, 53
Rural development, 13, 14, 27, 51, 55, 69
Rural Development Program, 27
Rural poverty, 67
Rural social movements, 76

S
Satisfactory, 98, 99
Self-creation, 104, 105
Small Organic Producers Forum in Zimbabwe (ZIMSOFF), 77
Small producers, 42, 49, 50, 76, 83, 93
Social dissatisfaction, 32
Social metabolism, 109
Social movements, 75, 131
Social processes, agroecology
accumulation by dispossession, 78
agro-extractivism, 78
Cuba's modernization, 83
extractivism, 78
food sovereignty, 78
grassroots organization, 78
green revolution, 78
horizontal socialization, 80
learning methodologies, 79
liberation theology, 79, 80
modernizing development project, 78
peasant-to-peasant, 80, 81
promoters, 81
social movements, 79
soil-conservation practices, 81
subaltern social sectors, 85
training camps, 84
Soil degradation, 30
"Soy republic", 73
Spiritual transformations, 127, 132
Sustainable intensification, 31, 125
Synthetic fertilizers, 30
Systematic land acquisitions, 39

T
Technique, 112–114
Technological biological power, 9
Technology, 112–114
Technology-led development program, 24
Temporality, 11
Territorial control, 35, 53, 58, 66, 77
Territorial hoarding, 46, 47, 61
Territorial rents, 36, 57
Territory, 46
deterritorialization, 47
phenomenological reading, 46
The second law of thermodynamics
agro-biotechnology, 109
ecosystemic organizations, 108
entropy, 106
equilibrium, 107
green plants, 106
humans' immense creativity, 110
industrial agriculture, 110
industrial agroextractivism, 109
living organisms, 108
non-equilibrium, 107
self-organizing processes, 106
solar and human energy, 110
tropical forest, 107
"waste" and "loss", 106
Time, 11
Tropical forest, 107

U
Urban industrial production, 5

W
"Wheat for Oil" program, 27
World Economic Forum's New Vision for Agriculture, 52

World Food Program (WFP), 26
World Trade Organization (WTO), 28
World Wildlife Fund (WWF), 29
Worth mentioning, 73

Z
Zero Budget Spiritual Agriculture, 92
Zero Budget Spiritual Farming effort in
India, 84

The manufacturer's authorised representative in the EU is Springer Nature Customer Service Centre GmbH, Europaplatz 3, 69115 Heidelberg, Germany. If you have any concerns regarding our products, please contact ProductSafety@springernature.com

Printed and bound by CPI Group (UK) Ltd, Croydon, CR0 4YY
30/09/2026
02225470-0015